Graduate Education in the 21st Century: Meeting the Challenge

Graduate Education in the 21st Century: Meeting the Challenge

Edited by

Susanne Huttner
Joa Rubaloff

Proceedings
Wednesday, October 2, 1996
University of California, San Francisco, Laurel Heights Campus

Sponsored by the Executive Committee
of the University of California
Systemwide Biotechnology Research and Education Program

Published by:
University of California
Systemwide Biotechnology Research
and Education Program
345 Giannini Hall, #3100
University of California, Berkeley
Berkeley, California 94720-3100 USA

Printed in the United States of America

Library of Congress Catalog Card Number: 97-81390
ISBN 0-9662665-0-1

This book is dedicated to the memory of Michael Cronin who was an advocate for advancing life science graduate education in a time of great change.

Contributors

Bruce Alberts
President, National
Academy of Sciences

Richard Atkinson
President,
University of California

Richard Attiyeh
Vice Chancellor
for Academic Affairs,
UC San Diego

M.J. Finley Austin
President, Association
for Science Professionals

John Barton
George E. Osborne Professor
of Law, Stanford University

Michael Cronin
Senior Scientist, Endocrinology
Genentech

A. Stephen Dahms
Professor of Chemistry,
San Diego State University

David Gollaher
CEO,
California Healthcare Institute

M.R.C. Greenwood
Chancellor, UC Santa Cruz

Susan Hackwood
Chairman, California Council
on Science and Technology

Craig Heller
Associate Dean for Research,
Stanford University

(More)

Contributors
(Continued)

Donald Kennedy
President Emeritus,
Bing Professor
of Environmental Science,
Stanford University

C. Judson King
Provost,
University of California

Maresi Nerad
Director of Graduate Research,
Graduate Division,
UC Berkeley

Elizabeth Neufeld
Chairman, Department
of Biological Chemistry,
UC Los Angeles

Edward Penhoet
CEO, Chiron Corporation

David Perlman
Science Editor,
San Francisco Chronicle

Kenneth Shine
President,
Institute of Medicine

Keith Yamamoto
Chairman,
Department of Cellular
and Molecular Pharmacology,
UC San Francisco

Editors' Introduction

The United States has a long and distinguished record of advancing the frontiers of research in the life sciences by sustaining excellence in the knowledge and skills of its researchers. When the National Research Act of 1974 (P.L. 93-348) established the National Research Service Awards (NRSA) for student research fellows and institutional training programs, it created a forceful and coherent strategy for recruiting the best young minds for productive careers studying "the physical and mental disease and impairments of man." Thousands of promising young scientists have since entered the basic biomedical, clinical, and behavioral sciences to solve critical health problems, ranging from heart disease to AIDS and from nutrition to substance abuse. They have participated in and even driven the "biological revolution" that is giving us remarkable new insights to the molecular and genetic basis of health and disease.

In the last quarter century, bioscience research and graduate education have changed. A 1994 report from the National Research Council ("Meeting the Nation's Needs for Biomedical and Behavioral Scientists") observed:

> The continuing need for highly trained specialists to conduct research to meet the health needs of the country is as great today as it was 20 years ago. However, because of changes in patterns of research funding and the structure of the marketplace, the nature of this need has changed somewhat in recent years. Today there is a greater demand than in the past for talented health scientists to provide leadership in industrial research settings, in federal government laboratories, and in hospitals and clinics.

In 1995, the Committee on Science, Engineering, and Public Policy (COSEPUP) of the National Academy of Sciences issued a seminal report, entitled "Reshaping the Graduate Education of Scientists and Engineers," which stated:

> A broader concern is that we have not, as a nation, paid adequate attention to the function of the graduate schools in meeting the country's varied needs for scientists and engineers. There is no clear human-resources policy for advanced scientists and engineers, so their education is largely a byproduct of policies that support research. The simplifying assumption has apparently been that the primary mission of graduate programs is to produce the next generation of academic researchers. In view of the broad range of ways in which scientists and engineers contribute to national needs, it is time to review how they are educated to do so.

Nowhere in the nation are these changes or the needs for a "human resources" policy more apparent than in California. California is the birthplace of modern biotechnology and the site of its transformation into an entirely new high technology business sector – commercial biotechnology – mediated by relationships between entrepreneurs and scientists. As the number of California firms has quickly grown (reaching one-third of all U.S. biotechnology companies), the state has established itself as world leader in harnessing a new international economic development vector. With that success has come an expansive array of human resource needs and opportunities that are largely dependent upon the graduate education process.

The development of commercial biotechnology in California reveals a certain natural synergy between the private sector and publicly funded graduate education and research. California universities, with their outstanding basic research and graduate education programs, have provided the essential foundation on which new products, technologies, companies, and jobs have been created. They form one of the three essential pillars of the biotechnology economy. The other two are entrepreneurial excellence and strong capital markets. The University of California system, Stanford University, Cal Tech, Scripps Institution, and other California research institutions have proved to be magnets for private investment in California. Start-up firms are tightly clustered around the state's campuses. For example, 1 in 3 U.S.

biotechnology companies is located within 35 miles of a University of California campus.

Success evolved over several decades. In 1960, the California Master Plan for Higher Education created a strategic plan for excellence in research and graduate education, as well as in undergraduate education. It specified the University of California system, among various higher education systems, as the state's research arm and created a blueprint for focusing resources to build a strong science and engineering faculty. Over the past thirty-six years, the Master Plan and federal research funding (largely from the National Institutes of Health) have worked synergistically, enabling California to create a unique capacity for innovation that has driven the state economy.

This capacity continues to be embodied in *people* -- faculty, students, and other researchers working in federally funded projects in academic laboratories, and scientists and engineers working in the private sector. More than any other single factor, *people* drive biotechnology. The knowledge and skills of people fuel innovation and provide leadership. For example, in California, 1 in 5 firms was founded by a University of California scientist and more than 85% of firms employ UC alumni. Viewed this way, commercial biotechnology can be counted as a product of more than 40 years of federal and state support for research and training in the biomedical sciences.

California's success has also revealed challenges that graduate education institutions must address if the scientific workforce is to keep pace with opportunities and needs in the coming decades. First, as advances in the biosciences themselves have evolved, the nature of research and the research environment in which graduate training takes place have changed dramatically. Consider just two cases in point -- computational biology and biomaterials. These fields have advanced through new, broadly cross-disciplinary approaches that wed life sciences such as genetics, biochemistry and molecular biology, with physical sciences, like physics, chemistry, computer sciences, and mathematics. This has created new opportunities for advancing research and applications. It has also created new needs for highly skilled personnel who are trained in different ways than we have trained graduate students in the past.

Second, there are significant emerging needs for scientific personnel in non-traditional fields outside of academic research. It is not at all unusual in California to find people with advanced biosciences degrees who are pursuing careers in business, finance, venture capital, law, regulatory affairs, and journalism. You also find them in teaching, basic research, clinical research, and R&D positions that are somewhat better aligned with conventional graduate training. Even in those career paths, however, training needs are evolving. Diverse employment opportunities place pressure on the graduate education process, demanding greater flexibility in the breadth, organization, and goals of the graduate curriculum. They also create a challenge of a different kind. Faculty are faced with graduate students who increasingly wish to pursue careers that diverge from the traditional academic path. Whether faculty approve and support student goals, or disapprove and discourage them, has significant impact on student choice.

Third, when this conference was held in October 1996, another challenge was emerging. The future of federal support for basic research and graduate training was uncertain. Congress was struggling to balance a federal budget plagued by a serious deficit, and political plans suggested that federal support for bioscience research and training be cut by as much as 25% by the year 2006. This came at a time when the California economy was finally starting to move on the upswing after a long and serious recession. It was a time when advances in biosciences were fueling rapid development of commercial biotechnology in the state and the nation, and when the biosciences were poised to enhance other important sectors of the economy, including agriculture and natural resources. When you consider that California universities account for 15% of all NIH extramural funding, it is evident that the proposed budget cuts would have serious consequences for the state's biotechnology enterprise -- both in academia and industry.

These consequences would inevitably reverberate through the national biotechnology economy.

These challenges and others defined a common purpose for the speakers at the conference held at the University of California at San Francisco on October 2, 1996. The discussions could have readily focused on how we can do more with less in light of proposed federal cuts. There was, at the time, a real possibility that we would be forced to do so. Wisely, the conferees took a different course, focusing on what value the nation and our state have gained from federal investment in research and graduate education -- and what must be done to preserve and sustain that value in a climate of change.

The articles in this volume are based on speeches made by a distinguished panel of university, government, and industry leaders. Those national leaders who contributed to Section One (Atkinson, Alberts, Shine, Kennedy, King, and Penhoet) clearly articulate the critical juncture in research and graduate education at which the nation and its graduate schools find themselves. The emerging consensus is that we must move away from the conventional wisdom that says that successful graduate students will eventually pursue the same careers as their faculty mentors. We must consider how to make graduate education more broadly useful to the expanding range of communities that depend upon the product of that education.

One article (Kennedy) emphasizes that faculty and graduate education institutions must challenge the attitudes, expectations and reward systems that set boundary limits on the graduate experience. Another (Penhoet) argues that, as graduate institutions try to adapt to emerging needs, they must resist the temptation to replace an effective educational experience that builds broad strengths in the fundamentals with one focused on highly specialized training that meets short term workforce needs. These perspectives echoed throughout the conference.

Section Two presents views on the role, impact, and utility of life science graduate education and research training. Discussion ranges from current and possible future workforce needs to increasing calls from graduate students and postdocs for greater flexibility and accountability in the training process. This section also considers the increasing role and impact of life scientists in law and public policy.

Section Three focuses on the experiences and perspectives of the dean, the department chair, and the faculty -- people who define the nature and scope of graduate education at research institutions. It also identifies an additional driving force for change in graduate education: increasing competition among graduate schools for a shrinking pool of outstanding graduate student candidates.

As a final comment, it was repeatedly acknowledged that the discussions at this conference and in national reports on graduate education all suffer a common shortcoming: a lack of accurate data. What are the national needs for biomedical scientists? Where are they needed? What kinds of education and training is required? How well is the current graduate education system doing? How are our graduates employed, and to what extent are they productively using their graduate education?

The strength and effectiveness of any institutional or national plan relies upon sound data. Research undertaken by Maresi Nerad (this volume) will provide important insights. Statistics on employment patterns provided by the National Science Foundation, scientific societies, trainee associations, and trade associations are also helpful, but insufficient. A deeper view is needed. An institutional study undertaken by the University of California Biotechnology Research and Education Program has been taking such a deeper view. It identifies and quantifies the linkages -- that is, the people that form those linkages -- between research and graduate education at the University of California and commercial biotechnology firms in the state. Some of the early data, summarized in Table 1, shows clearly that the California biotechnology economy depends heavily upon basic research and

Table 1

The University of California Connection in Commercial Biotechnology

1. 1 in 3 U.S. biotechnology firms is located within 35 miles of a UC campus, accounting for nearly half of U.S. biotechnology jobs and sales revenues
2. 1 in 5 California biotechnology firms was founded by people with advanced UC degrees or by UC faculty
3. 6 of the 10 top-selling biotechnology drugs stem, in part, from UC research
4. 85 percent of California biotechnology firms employ UC alumni

Source: The Critical Linkages Project, University of California Biotechnology Research and Education Program.

graduate education at the University's nine campuses. The study also shows that life scientists are being called upon to play a wide array of important roles, an increasing number of which are not considered in graduate education.

The study shows that individuals, with creativity and personal initiative, have bypassed conventional academic or industry research careers. Many have forged new paths by adapting the scientific knowledge and critical thinking skills that they gained in bioscience Ph.D. programs and in postdoctoral work. By being flexible and open-minded in defining opportunities, they have discovered productive and satisfying employment in business management, law, regulatory affairs, and other divisions of biotechnology companies. Others have taken positions in government, journalism, and teaching. Some have gone back to school to gain an additional advanced degree, but many have entered new professions without them.

The study underscores the need for flexibility and for rewarding personal initiative in graduate education. As M.R.C. Greenwood says (this volume): "Universities are going to become knowledge centers that provide lifelong learning. Part of what we have to teach our graduate students is how to continue to access that knowledge and how to be comfortable with new demands."

The conference reported on in this volume was hosted by the Executive Committee of the University of California's Systemwide Biotechnology Research and Education Program. The Biotechnology Program supports interdisciplinary basic research and training at the University's nine campuses, three affiliated National Laboratories, and the Agriculture Experiment Station. The conference theme was directly relevant to the program's mission and goals. It grew out of discussions that we have been having over the past year about the future of graduate training in a changeable world. It has served as a launching point for discussions involving faculty, students, and administrators in life sciences departments throughout the University of California system.

The conference topics and agenda were developed by a graduate student (Rubaloff) from the University of California at San Francisco, who worked as an intern at the Biotechnology Program. Its overall focus is optimistic, but tempered by a healthy dose of skepticism common among graduate students and postdoctoral trainees, today. We both hope that it will help trainees and faculty to work in concert to evolve life science graduate education to meet national priorities.

Susanne Huttner
Director
Systemwide Biotechnology
Research and Education Program
University of California,
Berkeley

Joa Rubaloff
Graduate Student
School of Pharmacy
University of California,
San Francisco

Acknowledgments

We gratefully acknowledge the superb technical editing of this volume by Scott LaPierre of the Systemwide Biotechnology Research and Education Program. We also acknowledge the efficient organization of this Conference by Dee Minor of the Systemwide Biotechnology Research and Education Program at the University of California.

We thank the distinguished speakers for their excellent contributions to the conference and this volume -- and to the historic debate on the future course of graduate education in the life sciences.

Contents

Section I

California and the Nation: Issues and Needs

C. Judson King

Graduate education in the biological sciences has been and will continue to be centered about the Ph.D. degree. The needs for and the uses of the Ph.D. have been subjects of much study and discussion in recent years. There is, for example, a well-known Committee on Science, Engineering and Public Policy (COSEPUP) report on the subject of the future of the Ph.D. which calls for some broadening of the degree and recognition that there will be multiple consumers for the degree. I think this is very right. There is, I believe, considerable demand and will be even more demand for the Ph.D. in biological sciences in industry.

The Ph.D. can be classified as a "creativity degree", or a generation of new knowledge, or a generation of new approaches degree. There is much need for all those things within industry, including particularly the biotech industry. There is also a need for people in that industry who can work very closely with university faculty and graduate students, and the Ph.D. is an avenue towards that. The fact that there is a substantial industrial demand for the Ph.D. is actually no surprise to somebody who's a chemical engineer, like myself. The statistics for chemical engineering nationally indicate that about 10 to 15% of Ph.D.s go to universities upon graduation, and the rest go to other places. By far, the largest job market is not in universities.

The second point I'd like to make is that the subject discussed here, although the title centers on one discipline, is inherently multi-disciplinary. Biology cannot stand by itself today. It has very significant overlaps with other sciences. Examples are chemistry and physics, but obviously there are others, too,

---

C. Judson King is Provost of the University of California. . This article is based on remarks delivered at the October 2, 1996 symposium, "Meeting the Challenge: Graduate Education in the Biological Sciences in the 21st Century" held at the University of California, San Francisco.

including very important overlaps within engineering fields. There isn't a biological engineering program that attempts to be an accredited degree, but there is much in engineering that serves biology. I believe that as the years go on, we will find all engineering fields turning more towards biological problems and applications. With regard to multi-disciplinary approaches, I can't resist citing an example from my own field. My area within chemical engineering has been separation processes. Biotechnology certainly arrives with very significant separation processes. A recollection I have from the past is the day I was called in as a consultant by a major pharmaceutical company -- not on this coast -- and was given a problem or two to contend with. The problem stemmed from the fact that this company had two very large buildings in its research area. One building housed the creation, the genesis, of the company's products -- the reaction, if you will. It housed more synthetic chemistry than fermentation or bioengineering, but that will change as the years go on. Large pipes flowed in one direction across the street to another building, the separations building. The separations engineer received all of this soup and tried to take it apart into the appropriate products, recyclables, and what have you. Clearly, these two endeavors need to be in the same building, at least intellectually. One of our challenges is to find ways to make the biologist more familiar with those things that will impinge upon the creation of the biological process and to get those engineers who work with the biologist more familiar with the biology, so that each can see what the other may have to contend with. Hopefully they can then talk with one another better.

So what does this mean for graduate education? There continues to be a need for a strong Ph.D. I would agree with the COSEPUP report that we need to broaden the content of that Ph.D., and I believe we need to find ways to reinforce the multi-disciplinary aspects. The traditional ways of doing that sort of thing are that one might get a masters degree in one area, follow that with a doctorate in another area, and thereby be jointly trained

The other way, of course, is through minors and other means of bringing disciplines together, including the use of graduate groups as we do within the University of California.

1 The Numbers Game and Graduate Education

Richard Atkinson

The National Science Board report, *Science and Engineering Indicators, 1996,*[1] has a new section this year, entitled "Science and Engineering Labor Market." It begins with the following statement: "The performance of the U.S. economy is the major determinant of current and future demand for scientists and engineers." I would argue that this statement represents a short-term perspective on the science and engineering labor market. Clearly the current economy determines the flow of taxes, company revenues, and the number of individuals who will be hired at any given time. A long-term perspective, however, would focus on the importance of science and engineering as a driver of the future economy; the investments made in R&D today will be a dominating factor in the level of economic growth experienced in the future.

For 50 years we have had a good understanding of the role of education as a driver of the economy, but only in the past 10 to 15 years have we begun to understand the impact of R&D on economic growth. When I was the Director of the National Science Foundation in the late 1970's, we had ample anecdotal evidence about the relationship between innovations like the transistor and this or that economic variable, but very few solid economic studies. We initiated a special program at NSF in economics focused on just that issue -- the relationship between investments in R&D and

---

[1] National Science Board, *Science and Engineering Indicators, 1996* (U.S. Government Printing Office, Washington, D.C.).

Richard Atkinson is the President of the University of California. This article is based on remarks delivered at the October 2, 1996 symposium, "Meeting the Challenge: Graduate Education in the Biological Sciences in the 21st Century" held at the University of California, San Francisco.

economic growth. A substantial literature has evolved in the interim, which in turn has led to a development in economics called "new growth theory." This work was nicely summarized in a recent report of the Council of Economic Advisers:[2] 50% of the growth in the American economy in the last 40 years has been due to investments in research and development. Obviously, the private sector is a major driver of that R&D, but federally funded research of universities also plays a key role. The report points out that when investments in university research increase, there is -- with an appropriate lag -- a corresponding increase in private-sector investments. There is a well-researched link between university-based research and industries' R&D efforts.

Investments in education and investments in R&D are overriding factors in economic growth. The nation faces tremendous problems -- deteriorating inner cities, homelessness, degradation of the environment, the prospect of a huge number of baby-boomers retiring with a far smaller workforce to support them in their retirement. How are we going to deal with these problems? There is only one way -- we must have substantial economic growth in this country. That means we are going to have to invest heavily in education and in R&D.

During the course of this book you will read about several initiatives by the University of California. One of these initiatives is our Industry-University Cooperative Research Program. The purpose of the program is to create jobs and protect California's high-tech leadership by forging research partnerships between UC and industry. I am very excited about the program and can report that it already is having an impact. The program will be given extra impetus by the R&D tax credit established through bipartisan legislation this year. The legislation doubles from 12% to 24% the state tax credit California industry will receive for the support of research at universities. These are both important investments in

[2] "Supporting Research and Development to Promote Economic Growth: The Federal Government's Role," The Council of Economic Advisers, October 1995.

California's future productivity and thus the state's ability to thrive in the next century.

Returning to the focus on graduate education. In 1990 I published an article in *Science* on the supply and demand for scientists and engineers, based on work done in 1988.[3] This article reported on an NSF study that I was involved in, much like the study that Bill Bowen and a colleague at Princeton were doing at about the same time.[4] Bill and I were both projecting a significant future shortfall of Ph.D.s. Bowen was looking at the humanities and the social sciences as well as the natural sciences and engineering. My paper was concerned only with the natural sciences and engineering, and excluded the social sciences. The study began with the year 1988 and projected the supply of Ph.D.s that would be trained in future years. That projection was made on the assumption that a certain percent of undergraduate students would go on for Ph.D.s, and thus was based on the demographics of the 22-year-old population. If you look back over the past 20 years, the proportion of 22-year-olds who eventually earned a Ph.D. in science and engineering is remarkably stable.

Added to that was the assumption that the number of foreign students taking Ph.D.s in the United States in future years would remain at the 1988 level. In 1988 we had a large number of foreign students taking Ph.D.s, and the assumption was that this number wouldn't increase significantly. A further assumption was that 50% of the foreign students who earned Ph.D.s in the United States would stay in the United States. And these assumptions led to the wiggly curve on the graph labeled "Supply of Ph.D.s" (see Figure 1, next page). You can see on the far left the actual number of Ph.D.s produced in 1988. Supply was projected through the year 2010.

---

[3] "Supply and Demand for Scientists and Engineers: A National Crisis in the Making," *Science*, Vol. 248, 27 April 1990.

[4] W.G. Bowen and J.A. Sosa, *Prospects for Faculty in the Arts and Sciences*, (Princeton University Press, Princeton, N.J., 1989).

Figure 1

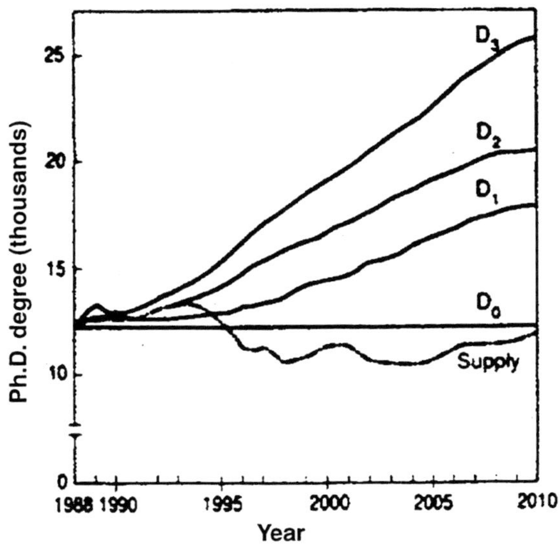

Supply and demand projected to the year 2010 for Ph.D.'s in the natural sciences and engineering. Four different demand scenarios are indicated by the D_0, D_1, D_2, and D_3 curves.

Reprinted with permission from *Science,* 27 April, 1990, Vol. 248, p. 429. Copyright 1990 American Association for the Advancement of Science.

The D_0 curve was based on the assumption that the future demand for Ph.D.s would remain constant. That is, whatever the demand was in 1988, that demand would stay the same out to the year 2010. Note that for the D_0 demand curve, it is not until the late 1990's that an undersupply of Ph.D.s begins to occur.

But the constant D_0 scenario seemed highly unlikely for at least three reasons. First, yearly replacements due to retirements and deaths were expected to increase over the next two decades. Second, we considered it almost certain that college and university enrollments would increase in the late 1990's with the expanding college-age population, necessitating an increase in the number of faculty hired. Third, we assumed that if federal and private investments in R&D continued to grow at even moderate rates, the number of new Ph.D.s required by industry would be well above the 1988 level. These three factors generate three cumulative demand scenarios labeled D_1, D_2, and D_3.

We knew the demographics of the workforce in 1988 and the age distribution of Ph.D.s in that workforce, and therefore it was fairly easy to predict the expected increase in the number of retirements. If one assumed that for every Ph.D. who retired, a replacement would enter the workforce, the result is the D_1 curve. If you compare the D_1 curve with the supply curve, somewhere after the year 2000 you begin to see a significant divergence; an increase in Ph.D. production would have to occur in order to have an adequate supply.

Another factor was the number of undergraduates enrolled in universities. We could predict with assurance that there would be a significant increase in the number of college-age students by the late '90's. If one assumed that the ratio of faculty to students would be maintained at the 1988 level, then one would add to the D_1 curve and predict a demand for Ph.D.s represented by the D_2 curve. Finally, assuming there would be growth in the number of Ph.D.s required in the non-academic workforce, it seemed sensible to add a growth factor of 4% which cumulated to the D_3 curve. After all, the private-sector workforce was expected to expand, and one would expect a correlated increase in the need for Ph.D.s.[5]

These four demand curves represented projections based on a well-defined set of assumptions. And depending on which curve you believed, you could get quite exercised about the projected shortfall. Once I published the 1990 paper, and once Bill Bowen published his work, considerable unhappiness ensued in the academic world as the job market began to deteriorate. The fact is, however, that both sets of projections did not identify significant shortfalls in the supply of Ph.D.s until the late '90's.

What has happened since the publication of the 1990 *Science* paper? One unanticipated factor was the end of the Cold War, with

---

[5] To save time, I have provided a very cursory explanation of these projections. The original paper should be consulted for a more detailed account.

the resulting cutbacks in defense spending. Another factor was that the number of foreign students taking Ph.D.s in the United States did not remain constant, as we assumed, but instead has grown at a significant rate. And in 1988 I thought a greater proportion of foreign students would choose to return to their country of origin than had been the case in the past, because those countries were becoming more competitive and more attractive for young Ph.D.s. It turns out that the proportion that remained in the United States did not decline, but rather increased.

What else has changed? Colleges and universities across the nation haven't yet experienced the kind of increase in enrollments that will be coming. Further, the student-faculty ratio has not remained constant, but rather has deteriorated. For example, at the University of California, it has changed from a ratio of about 14:1 up to almost 19:1. I hope the student-faculty ratio will return to more favorable levels in the future, but for the moment it is a clear indicator that the nation is investing fewer resources in educating college students.

Another factor was the Immigration Act of 1990. This act specified categories of individuals seeking to immigrate who had particular skills and gave them added consideration. As a result of this legislation, a large number of foreign-trained Ph.D.s have entered the U.S. labor force. Twenty-three percent of the Ph.D.s employed in the United States today were born in another country.[6] We now have the highest percent in history of foreign-born Ph.D.s in the United States workforce.

So the question is: Is there an oversupply of Ph.D.s? Several weeks ago I read in the *New York Times* an article[7] by a professor of history about the serious employment problems facing Ph.D.s, particularly in the humanities, where the placement rate for new Ph.D.s in history and English, for example, is less than 50%. It was a wrenching article. Without question, there are disturbing

[6] National Science Board, *Science & Engineering Indicators*, 1996.

[7] Louis Menand, "How to Make a Ph.D. Matter," *New York Times Magazine*, September 22, 1996.

problems in the humanities. Are there problems in science and engineering? That depends on whom you talk to. In physics, there is no doubt that we have a problem. But consider a field like engineering. If you are a Ph.D. electrical engineer or a computer scientist, there is an oversupply of jobs. In mechanical engineering, on the other hand, there is a shortage of jobs.

In recent years, the Science Indicators Report has included a new measure designated "involuntary/outside of field," meaning Ph.D.s who cannot find appropriate work and have been forced to work outside of their area of expertise. The fields of geophysics, physics, and mechanical engineering are the three highest in terms of the percent of individuals who fall into the involuntary/outside of field category. The figure is 7.7% for geophysics, and about 6.5% each for astronomy, physics, and mechanical engineering. For a few other fields: biological science, 2.1%; computer science, 1.4%; and chemistry, 3.5%.

The point is that Ph.D. employment is very much a field-by-field issue. A close friend of mine took a Ph.D. in astronomy -- a first rate degree and a first-rate talent. He now runs a software company on the peninsula. You can imagine his history: as an astronomer he did a great deal of work on instrumentation, which involved sophisticated computer programming. He tried for many years to land a regular faculty position, but eventually gave up. He then started his own software company and has done spectacularly well. Is he inappropriately matched to his Ph.D. program? Was it a mistake for the United States to invest in his training in astronomy? I think not. The biotech business would not be thriving in California if we did not have a steady flow of Ph.D.s from our universities.

Despite problems in a number of fields, therefore, it is disturbing to hear some people make the blanket statement that we are training too many Ph.D.s. Certainly physicists and humanists will resonate with that notion. But to jump to the conclusion that the nation faces an across-the-board oversupply of scientists and engineers is inaccurate and misleading. This doesn't mean we can't improve the preparation of Ph.D. students. Bruce Alberts writes about the National Research Council study on reshaping graduate education in science and engineering. I am enthusiastic about their recommendation to reduce time to degree. I also support the idea that the training of Ph.D.s should be more versatile so that they have greater opportunities in the job market. And the Council's recommendation for a national employment database for science and engineering Ph.D.s should be an immediate priority. Such a database would be invaluable to faculty advisors and to students as they plan their future. Further, if such a database were accessible on the Internet, we would quickly realize that the information we are currently collecting is not adequate. We'd begin to expand and refine our database and have more relevant information. So I support the Council's recommendations, particularly as a way of matching more closely societal needs with the training of Ph.D.s in various subfields of science and engineering.

The University of California is a major player in graduate education. We produce about 10% of the nation's Ph.D.s. Until this year, we have had a formula-driven budgeting process for graduate enrollments that makes little sense in the current environment. To greatly simplify, from about 1960 until this year, the number of doctoral students in a given discipline was principally determined by the number of undergraduates in that discipline. A large number of psychology undergraduates translated into a large number of psychology graduate students. The formula wasn't quite that simple-minded -- and it did take account of field-to-field differences -- but that was the basic idea. With much discussion among the faculty and little public fanfare, we have changed our budgeting process. The change takes effect this year. No longer will we tie the number of graduate students to undergraduate enrollment, field by field. We now have a budget process in which departments will not lose budgetary support if they cut back in their graduate enrollments. Until last year, departments had to have large numbers of graduate students in order to receive the full set of rewards that the system had to offer. We are now changing our budgetary system so that the number of

Ph.D. students in a department is driven more heavily by the job market and employment opportunities.

The topic of this symposium is biological sciences, and that of course involves medical education. It has become clear that we must re-think the nature of medical education -- how many students we educate, how many go in different specialties, the nature of that education, and related issues. We have established a commission whose charge is described in a document titled, "The University of California Commission on the Future of Medical Education." Dr. Tom Langfitt, who is the head of the Glenmede Trust in Philadelphia, is chairman of the commission. The individual who is organizing the work of the commission is with us today -- Professor Charles Wilson, a distinguished member of the faculty here at UC San Francisco. There will be a number of working papers from the commission and a final report on how we should rethink medical education in the state of California.

Let me conclude by saying that the training of Ph.D.s to meet the nation's needs is one of the most important questions facing higher education, now and into the twenty-first century. We must be very careful about how we think about graduate education and marketplace -- taking into account both short-term and long-term perspectives. We would do ourselves and the nation a disservice if we came to a blanket generalization that our research universities are producing too many Ph.D.s. The problem is too complex and too important to the nation's future to yield to simple-minded solutions.

2 The Forces for Change

Bruce Alberts

I have been away from San Francisco for a little over three years now in Washington, which makes me a bit rusty on some of the issues being discussed here. Although I graduated my last Ph.D. student only a few months ago, I have been out of contact with most of the graduate students at UCSF recently. But having spent 25 years deeply immersed in these issues, I hope that I can still retain some credibility in discussing graduate education today. Of course, in Washington one gains the advantage of a different perspective. One sees how the world outside of the universities feels about those of us at universities. This perspective perhaps makes one think more broadly than one can when immersed deeply within the university itself.

Let me start by discussing the National Academy of Sciences and the National Research Council, because we have been becoming increasingly active in the kind of issues that we are discussing here. The president of the Institute of Medicine, Kenneth Shine, and I see each other every day and the two of us, together with the new president of the National Academy of Engineering, work closely on human resource issues.

The National Academy of Sciences is a unique organization, although there are about 80 academies of sciences in the world. We received a special charter from Abraham Lincoln in 1863, which I'd like to emphasize. The following is an excerpt from the National Academy's Congressional charter in 1863 that explains who we are and what we do:

Bruce Alberts is President of the National Acedemy of Sciences. This article is based on remarks delivered at the October 2, 1996 symposium, "Meeting the Challenge: Graduate Education in the Biological Sciences in the 21st Century" held at the University of California, San Francisco.

> The Academy shall, whenever called upon by any department of the government, investigate, examine and report upon any subject of science or art...but the Academy shall receive no compensation whatsoever for any services to the government of the United States.

We were allowed to form an honorary association of the country's best scientists. However, there was a price to pay. Whenever called upon by any department of the government, we were "to investigate, examine and report upon any subject in science and art" (art meaning technology at that time), and here's the catch -- "but the Academy of Sciences shall receive no compensation whatsoever for any services to the government of the United States."

This charter started a long volunteer tradition in which scientists and many other people donate their time to advise the government. Today, with the subsequent founding of the Institute of Medicine and the National Academy of Engineering, the three institutions work together to oversee an expanded "academy," known as the National Research Council (NRC; see Table 1). Although we don't pay the people who are volunteers on our committees, like universities, we do charge our costs to the government for the services that we provide. At any one time we have something like 6,000 active volunteers, and servicing the committees requires about 1,100 staff members, most of them in Washington, D.C.

We publish about 200 reports a year. One of these reports has played a major part in stimulating the discussion that we are involved in today. This report, called "Reshaping the Graduate Education of Scientists and Engineers," was produced by our Committee on Science, Engineering and Public Policy (COSEPUP). One of its major recommendations was that we should greatly increase the career advice and exposure that we give to graduate students, emphasizing the many different ways in which they can use their scientific talents in the future. That is, we need to

Table 1

The National Research Council

- Founded in 1916
- Operated by the National Academy Complex (National Academy of Sciences, National Academy of Engineering, and the Institute of Medicine)
- 6,000 volunteers in any year
- 1,100 employees (60% program staff)
- Located in Washington, DC, satellites in Irvine, California and Woods Hole, Massachusetts
- Publishes 200 reports per year, all available at www.nas.edu

stress that there are many more good career options in addition to the traditional goal of becoming a professor.

Because we took this recommendation very seriously, we next developed the booklet "Careers in Science and Engineering -- A Student Planning Guide to Graduate School and Beyond." This publication is designed to help students think broadly about their careers, both before they go to graduate school and while they're in graduate school. Associated with this Guide is a still-developing electronic Web site -- a career planning center for beginning scientists and engineers, which connects them to information about various careers, both research careers and so-called "alternative" careers. In addition, it has the aim of connecting them to real human beings who can give them advice on whatever they're interested in through e-mail.

Both this site and the full text of the Career Guide and 1,000 other reports are available from our home page at www.nas.edu. I want to make an important point here, which is that the president of the Academy has one major prerogative: I have the final say on how our home page is organized. At the very right-hand top of our home page screen is a button called "For beginning scientists and engineers," which emphasizes how strongly we feel that the young scientists and engineers are a crucial resource that needs to be encouraged and developed by all of us in the Academy and in the community at large. There are some coming attractions in this series. Early next year we will have a successor to the Career Guide for students in the form of a booklet for faculty member mentors. We are advertising this as a resource for new faculty, pretending that established faculty already know everything about mentoring. But of course we hope that experienced faculty will also take this seriously. As with all of our products, this book will not only be published, but also put out on the Web for free access.

We are also in the midst of a detailed study of the career paths of young biological scientists. We are trying to answer this question -- where have all the young principal investigators gone? We did a study about a year ago, through a committee co-chaired by Shirley Tilghman and Torsten Wiesel, entitled "Young Investigators in the Biological Sciences." This study uncovered the remarkable fact that between 1985 and 1993, the number of "young" investigators (young being defined as under age 37), had declined by a factor of 2 as measured by those applying for independent research grants from the National Institutes of Health. Moreover, the number of those obtaining grants had declined by a factor of 3. At the same time, the total number of grants awarded had not changed, so that the whole population of independent investigators in the biological sciences had aged enormously. What's going on? Nobody knows for certain, but our new committee is going to find out.

I recently spent a few days in Mexico. It is always very useful to get outside your normal world and compare it with somewhere else. There is an Academy of Sciences in Mexico, and we have a partnership with them that has been very productive. For example, we recently carried out a major study on the water supply in Mexico City, using a joint US-Mexican committee, which helped their government appreciate the value of science in political decision making. Mexico City sits on an enormous aquifer, but they're going to run out of water if they don't do something intelligent soon, which is terribly serious for a city that has 20 million people in it.

We have also been involved with the Mexican Academy of Sciences in organizing a review of all the science and engineering graduate programs at their major university, called UNAM. This is their National University with 260,000 students (only 130,000 of them are comparable to the University of California, because they run high schools, as well as undergraduate and graduate schools). UNAM is the major intellectual-resource development agency for Mexico. The Rector, José Sarukhan, is a distinguished scientist who's a foreign member of our Academy. He asked the U.S. and Mexican Academies to collaborate to do something that had never been done before in Mexico: set up independent visiting committees to evaluate UNAM's various graduate programs. These visiting committees, half American and half Mexican, wrote a very honest series of reports. Initially these reports were supposed to be confidential. But to increase their impact, Dr. Sarukhan decided to publish them in a book and release them publicly.

I was recently in Mexico helping to roll out the critiques of these graduate programs to the public. They have many different graduate programs that seem to have grown up helter-skelter, some good and some poor. And our report was quite truthful and direct about the need for change. In the public release I gave a short speech in which I attempted to deflect the defensiveness that I would have expected from their faculty with respect to protecting the status quo. I believe that the central points I made there are also quite relevant for us today.

First of all, the future strength and prosperity of a nation, be it the U.S. or Mexico, will depend on its future capacity in science and technology. This is an obvious truth for the next century, which is more clearly recognized by Japan and some other Asian countries than by our own government. In developing countries like Mexico, it is not likely to be widely recognized except by the scientists themselves.

The second point is that we must pay special attention to the young people who are interested in science and technology careers, because they represent the future of the country to a large extent. This means that we must encourage and recruit the most promising young people to become scientists and engineers. This is an issue for the pre-college and undergraduate level.

Third, we must educate these promising young people well in graduate school, so as to make them optimally useful to society. That requires a good sense on the part of faculty about what society needs from its scientists and engineers. Last but not least, we must ensure that we provide good career opportunities for as many of these young people as possible, preferably for all of them.

In the long run this production of "human capital," as it is called now in Washington, is the most important function of the science and engineering departments in Mexico; it is even more important than the research that the universities do. And at the risk of alienating some of my colleagues, if I was forced to make a choice of our most important function, I would say basically the same thing about the University of California. Even more important than the research that we do are the qualities of the people that we educate, because they represent the future.

I feel I can be critical of some of my colleagues because I am really being critical of myself. I have been a professor for 30 years, and at one time or another I have had nearly all the attitudes that I now find misguided. For example, I once thought that my only role was to produce future professors just like myself. During my ten years at Princeton (1966-1976), our whole attitude was that anybody who wanted to go to medical school or anything else, that was sort of a necessary by-product -- but we were really in the business of finding the best young people and making them into professors.

Also like many of my colleagues, I once thought that the world of the university should be preserved as an ivory tower -- keeping it away from the "impure" goals of business and industry, for example. In fact, I came to the University of California in 1976 at the very same time that Herb Boyer was starting Genentech, which I and many of my colleagues thought was some kind of immoral act. I also felt then the way that many of my colleagues still feel -- that contact with politicians conflicts with the scientific ethic which relies on truth and straightforward logic, since politics is less honest and less rational.

Finally, there is the even more extreme view that all work that has anything to do with the practical problems of our society is less prestigious and valuable than work that is unrelated to these problems. In my opinion, a real weakness of the academic community is that we tend to respect those studies that are pure and independent of the vagaries of the real world much more than we respect intellectual work that is motivated by our societal needs. At the National Research Council, one of our major functions is to get academics together with police chiefs, people educating children, etc. -- so as to help engage the expertise of the academic community with some of the major needs of this country.

Why have I changed my mind on all of these issues? First, the era of rapidly expanding university faculties that I and all my colleagues grew up with has ended. It clearly never was sustainable. In a steady-state population of professors, I should produce one tenured professor in my career, whereas I already have produced about 20. Clearly, these 20 cannot produce another 20. As the COSEPUP report from the Academies pointed out, in recent years a smaller and smaller percentage of our Ph.D. graduates have been entering academic careers. Today, even in biology, only a little more than a third of our Ph.D.s are entering academic careers. So that if professors continue to be like I used to be -- only valuing academic careers for their students -- we are dooming most of our students to think that they are failures. If we do that, then of course very soon we won't have the best, the brightest, and the most energetic young people coming in as students. So we don't have any choice, we have to change our behavior even if we only want to preserve our own enterprise.

The second reason that I have changed my mind has to do with being in Washington. Since I have been there, I have come to realize something that I think is hard for scientists to believe. It turns out that the great support that we had from the public and from government for publicly supported research in science was dependent to a very large extent on the Soviet threat. The Soviet Union had outstanding scientists, as Sputnik made us acutely aware, and this led to an enormous investment in our scientific enterprise. Now that the threat is gone, we need an energetic new effort from the scientific community to make our value to the nation obvious to everyone. This means that the "ivory tower" must be replaced by a completely different metaphor. The university must be viewed as a regional and national resource that is reaching out and accessible to all.

Now what does this mean for our science departments? It means that science departments must be supportive, concerned about, and in frequent contact with all those who use science in their careers -- not just other professors, but of course those in industry, science teachers, science journalists, members of Congress and their staffs, those who make the scientifically-based decisions in local, state and national governments, and so on.

We may be about to suffer from the lack of attention that we pay to these issues, because we are not well represented in Congressional offices or in many other places where we should be represented. The only way to represent us adequately is with people who understand what goes on in universities, who appreciate what research and education are and why they're important, and who therefore can be effective in the various centers of power in our country.

In short, we must make the value of our vigorous and exciting scientific enterprise in the universities obvious to all. The main barrier to change is the problem of nearly everything I deal with in Washington -- inertia. The status quo has many characteristics of a stable self-reinforcing system. Chemists would say that we are at equilibrium, with strong restoring forces opposing any change.

I think that much of the problem lies with our faculty, who know almost nothing about careers other than academic ones. If they don't know anything about other careers, there is no way they can be expected to value them. And if they don't value them, then that becomes obvious to all the students. The good news, however, is that scientists do believe in evidence when they see it. So once they get to know several outstanding people in other fields: outstanding science teachers at the pre-college level, outstanding people who work in Congressional offices on science policy issues, outstanding science journalists, etc., faculty will recognize the value of these careers for their students.

I believe that the strategy we need to pursue in our schools is a simple one. It needs to start with the opening up of our science and engineering departments. We need to bring back to the campus all those who once were science and engineering students and have left and done many other things with their education. Of course, if we select outstanding people, that will stimulate both the faculty and the students. I give a few examples here. Bring back science journalists so that you really get to know them and they get to know us. They need to get to feel at home in our science departments -- not that they're some strange person over there at the Chronicle that one tries to get to publish one's latest discovery. Instead, we should be working with them to try to help them in their jobs, while they are trying to help us be understood. This has many other advantages. It educates the faculty, and it provides career information for students who have those goals. Exactly the same thing applies to Congressional staff, precollege science teachers, and so on.

It is not very complicated. We simply need to change the way that we think about the standard pipeline (see Figure 1). I would like to present a new pipeline, one where entering science students are seen as a national resource. Our obligation is to get them to the places where they're needed in society. Places where they will be advocates for more rationality in decision making and of course

Figure 1

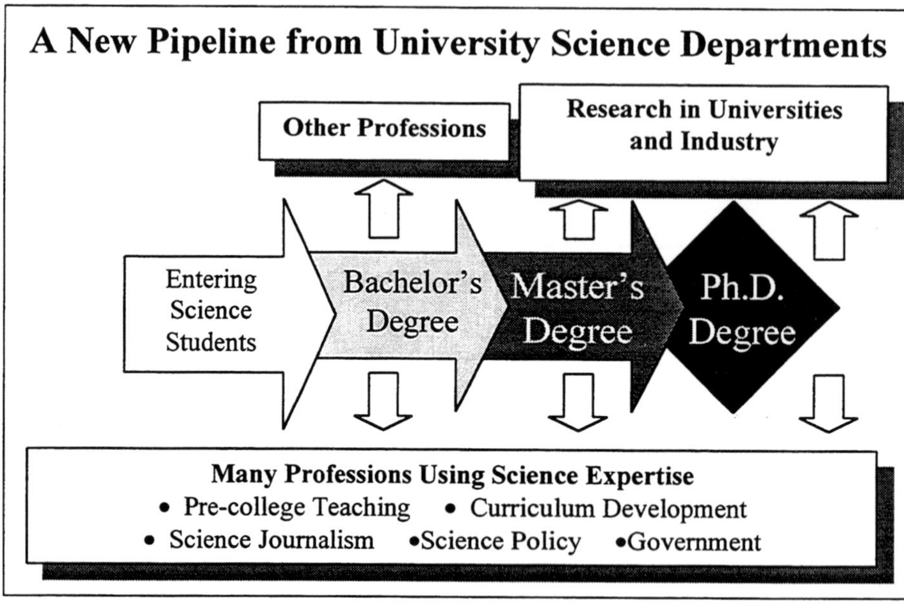

ultimately for science itself. And so we need many different career pathways that professors understand and encourage -- whether from the bachelor's degree, the masters degree, or the Ph.D.

There are really two different types of problems here and I have only discussed one of them so far. I have discussed the attitudes of the faculty, and I have discussed making career information and advice available to students. There is another major problem with the system. There are unnecessary practical barriers that now exist when a student decides that he or she wants to do something different -- like be a science journalist or be a pre-college science teacher, with no smooth pathway to get there in many cases. Let me take the example I know best, the one that represents perhaps the most extreme example of gridlock. There is an old front page of *Scientist* magazine (*Scientist*, September 20, 1993) which just happens to juxtapose two articles in a way that I found amusing. One headline is "Job Market for Researchers Will Remain Sluggish" and right below it, "National Science Foundation

Study Finds Many Teachers Unprepared for Instructing Children in the Sciences."

The Academy has been involved in promoting national science education standards. We are going to have on November 20th the announcement by the Secretary of Education of the results of the Third International Math and Science Study (TIMSS). Last time, US students were near the bottom in achievement. We don't know the results of this third international comparison yet, but we are all very sure that we are going to do poorly again. This time we won't be able to say that we did poorly because we tested only selected students in other countries. That was all controlled.

If we are going to bring science to the public, if we are going to prepare students so they can get reasonable jobs out of high school in manufacturing and other places -- rather than being forced to flip hamburgers at McDonald's -- we are going to have to address this issue head on. That means making science a core subject in every year of school starting in kindergarten, as it now is in England. And it means getting a large group of new people who really understand science as teachers in our schools. This is an opportunity for many young people interested in science to do something very productive with their lives, and it is a challenge to all of us to see if we can make it happen. But, we have these incredible barriers that science students must confront in order to get into this career. They must get certified to teach through State certification requirements that don't make any sense in many cases. Many of the courses that they are required to take are the result, in my opinion, of the education faculty wanting to get more faculty positions (FTE's) in their department -- therefore pushing for more course requirements, and so on (see Table 2).

We in science departments are also to blame. We make students leave the scientific community by telling them that "if you want to teach at the pre-college level (or even perhaps in community colleges), we are not going to have anything to do with you anymore." That is a real mistake. And it leaves these people isolated in the schools once they get there.

Table 2

Movement of Graduate Students and Postdocs Into Pre-College Teaching

Today's barriers:

- Must go back to education school to get certified; many courses are not worthwhile
- Must leave scientific community
- Students are isolated in schools

Tomorrow we must have:

- Direct movement into schools after appropriate introductory training, followed by mentoring
- Remain connected and supported by universities and the scientific community
- Community between teachers in schools, facilitated by special World Wide Web sites

So tomorrow we must have a different system that makes it possible for our students to think about careers in a more rational way. I like to advertise the programs at Harvard and MIT, whose science majors can get certified to be teachers as part of their regular undergraduate degree. That is not true most places. For example, I recently visited the University of Kansas. They have an exploding number of biology majors in their undergraduate programs. What are they all going to do? Well, a lot of them would like to be teachers. What do they have to do if they want to be a teacher? At the University of Kansas, you've got to go three more years to their education school to be certified after you get your B.A. What young, talented scientist is going to want to do that? This is a counterproductive situation, which is both bad for the students and bad for society.

The last thing I want to discuss is that the inertia problem extends far beyond the problem of getting our students into so-called "alternate careers" (I think we need a less pejorative description for these careers today). As Ken Shine implies in Chapter 3, there is also a problem within our own disciplines. Just to talk about my own world, fundamental cell biology, my students there tend to do the same kind of research I do after they graduate. And there is another whole field that comes largely out of medical schools, where students work on bone, or kidneys, or other tissues. Again these young scientists tend to do the same things that their professors did after they graduate. Yet there is a tremendous opportunity for new kinds of work at the interface between these two types of fields today. The students who learned fundamental cell biology can make an enormous contribution to medicine and to problems of human biology. But there is no real way to get them there. There are no good models. I know a very bright young scientist who is trying to make this transition, but he tends to get caught in limbo, stranded between two different types of scientific communities.

We have a strong barrier to getting new science started with young people. The attitude of the academic community I came from is that the free market is perfect; even though many of us don't feel it is perfect in other areas of American life, it is perfect in academia. The example we use over and over is the R01 independent investigator grant -- an unfettered grant request at the NIH with no guidance. We say "just leave it to the curiosity of the investigator," and I certainly believe in that. But that is only one extreme. The other extreme we use is the "war on cancer," where they made these incredible pie charts that tried to predict everything in advance; this is our image of planning. Partly because of that experience of the 1970's, we want to have nothing to do with planning whatsoever.

Figure 2

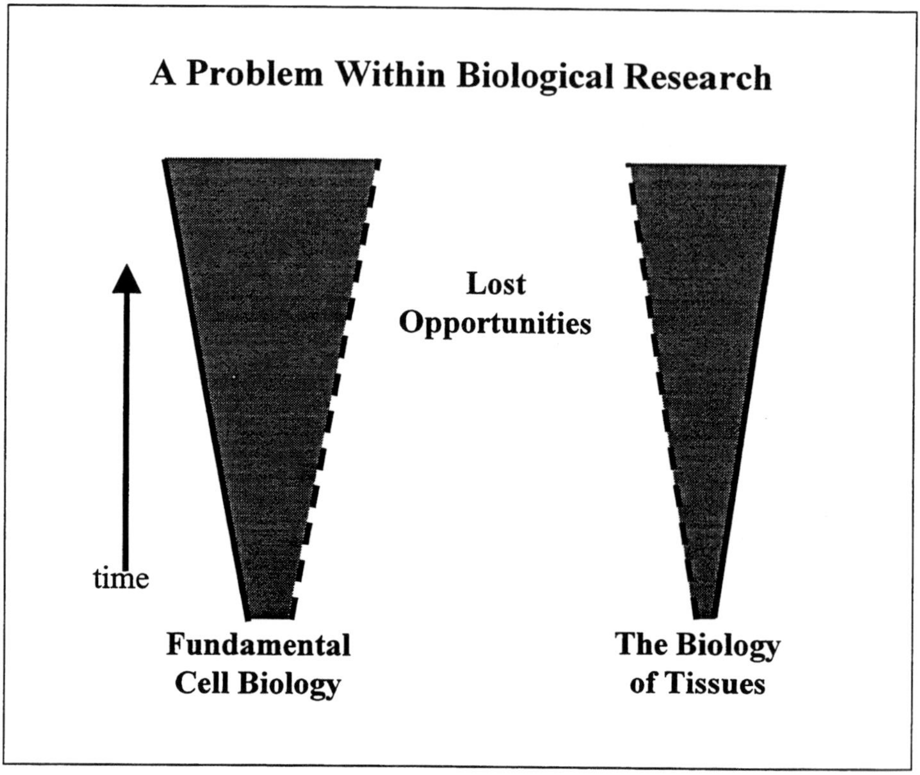

I think that there has to be an intermediate position that admits that intelligent planning is important and must be done, both in research and in scientific careers for young people. But this is going to take intelligent leadership. It is the role of the people in this room and others like us to exert that leadership. What does that mean? The only way you can lead in a university is with resources. Faculty will not follow directions; unlike in a corporation, they're not afraid of losing their jobs. So my experience at UCSF was that the skillful control of resources was the only lever that allowed deans to get needed things to happen. The control of resources in universities is important. You have to think about how you can use that control to drive productive change. It is also important at the national level. How can we design our federal support for science so that, in an intelligent and effective way, we exert the kind of direction that is needed on all the issues that I have discussed today? These are real challenges for us all. But the attitude that leadership is not important and that everything works fine in the free market is the path to ruin and despair.

3 Sustaining Excellence

Kenneth Shine

As a medical educator, it is of some interest when people worrying about graduate education argue about whether there are too many Ph.D.s. We have been arguing about whether there are too many doctors for a long time and, as you know, there is a concerted effort being made nationally to try to do something about the apparent oversupply of physicians.

The Israelis have dealt with the same issue for a long time. Large numbers of physicians go to Israel from Eastern Europe and around the world and they must find a variety of ways to support themselves. This calls to mind the wonderful story of the bus going up the hill to Jerusalem. When a woman collapsed, three physicians jumped up to give her CPR. The bus driver slammed on the brakes, stood up and announced, "My bus, my patient!"

I participated in the production of the Committee on Science, Engineering, and Public Policy (COSEPUP) report and I associate myself with the notion that there are clearly footprints of excess numbers of Ph.D.s in certain areas. Overall, our capacity to predict future workforce needs is flawed. It is very difficult to make long-term projections, but I will make the argument that in the life sciences, opportunities will be expanding. Moreover, any serious attempt to inhibit these developments would be counterproductive.

In any market, outcomes can change in a very dramatic way if people understand what the market is doing and what it needs. In

---

Kenneth Shine is President of the Institute of Medicine. This article is based on remarks delivered at the October 2, 1996 symposium, "Meeting the Challenge: Graduate Education in the Biological Sciences in the 21st Century" held at the University of California, San Francisco.

medicine, while it has been clear since 1978 that we needed more generalist physicians, the actual proportion fell. When the market began to pay more for generalist physicians and the jobs turned out to be there, medical students entered generalist subspecialities in increasing numbers. That trend is continuing. In the case of graduate students, the necessity for a national data base is critical to allow graduate students to make rational decisions about and what they want to do and where they want to do it. If accurate information is available, the right things will happen.

I do not accept the hypothesis articulated by many that federal funding of fundamental science will decrease in all areas. In the life sciences that will not happen -- and results of the last two NIH funding cycles support that conclusion. There must be a clear and well-articulated message that the life sciences are increasingly important. I shall comment on the nature of the life sciences and particularly on challenges for a great university like the University of California to do more at the interface between the academic health centers and the rest of the campus. I shall also discuss clinical research because it is critically important for both M.D.'s and Ph.D.s.

The first half of the 20th century was dominated by developments in the physical sciences. From Einstein's theory of relativity through atomic physics and radar, and from the space program to the need for military preparedness, the 20th century was built on physics, engineering, chemistry, and computer technology. That's not to say that there weren't important developments in the life sciences -- the double helix and biotechnology are among the most important. In the next century, however, many fields of the physical sciences and related areas will be important, but the biological and life sciences will offer new opportunities in almost every aspect of human endeavor: from agriculture, animal and fish husbandry, environmental palliation, to the production of compounds in the chemical industry by biological means; from genetic therapy to memory and the impact of the biological and behavioral sciences upon the human experience. The life sciences will be the leading sciences of the 21st century.

Opportunities for well-educated scientists in the biological sciences will be substantial, although relatively few of these will be in traditional academic roles. There has been an expansion of job opportunities in industry after a long period in which overall United States industrial investment in science was falling. Over the last two years, some industries have shown double digit increases in R&D. Ninety percent of that R&D is development. Less than 10% is in fundamental research, and of that 10% a major portion is called, "targeted basic research." It is clear that industrial investment is beginning to expand.

From my perspective, the message about the importance of the life sciences -- that space and defense were the principal goals of the 20th century federal science funding strategy but the quality of life will be the goal of science in the 21st century -- is a message that needs to be communicated. We need to communicate the concept that poverty, pollution, poor jobs and disease are every bit as dangerous as the evil empire. And we need to articulate that in a way that will encourage public expenditures in the life sciences.

It is clear that information systems will enhance the opportunities for scientists working in one area to be in rapid communication with those working in another, both domestically and internationally. New technology opportunities will develop which synergize the physical and biological sciences, not only in bioengineering but in such areas as the tertiary structure of proteins, the organization of cells, organs and organ systems, that move from the reductionist accomplishments of molecular biology to an integrated understanding of how organisms function. These developments will further enhance the role of life scientists who are trained in the physical and computational sciences. Major obstacles appear when the physicists are on the main campus and there are opportunities in the academic health center or in the biological sciences in another part of the campus to bring to bear the power of physics on the cell and the organ.

Similarly in the behavioral sciences. It is recognized that almost half of all illnesses in the United States, including those produced by alcohol, tobacco, sexual behavior, violence, accidents and drug abuse, increasingly require sophisticated research in the behavioral sciences further emphasizes the importance of university research to the life sciences. In many cases, the outstanding investigators in the behavioral and social sciences are on the general campus. Bridges must be built between researchers on the general campus and those at the academic health center.

There is no question that economic pressures on research institutions are producing both challenges and opportunities in this regard. For example, we are seeing an increase in consolidation of institutions -- this conference occurs in the shadow of potential changes at the University of California at San Francisco. I estimate that, annually, between $2.5 and $5 billion of clinical income to U.S. academic health centers cross-subsidizes research and education. This probably represents as much as 20% of all research money spent by academic health centers. Many of our faculties do not understand that patient care dollars subsidize not only clinical research, but also fund research in the basic science departments through the dean's tax and other pathways. The strategies which enhance both clinical research and basic science research are mutually interrelated.

The COSEPUP report focused on the notion that graduate students be prepared for flexibility in their approach to science, have the capacity to move from area to area, and be able to work in industry. Graduate students are also pivotal to the advancement of interdisciplinary science. They are, in many cases, both the glue and the grease that allows interdisciplinary research to take place. The opportunity to work with investigators from multiple disciplinary backgrounds not only allows exposure for the students, but also brings their sponsors together. The opportunity to have a graduate student working at the interface is critical.

Yet, there remains a fundamental problem in the way we fund our graduate students largely through NIH RO1-type grants which require that the student's work be related to the mentor's funded project. This phenomenon clearly inhibits interdisciplinary and multidisciplinary research. The relative amount of support available through training grants, teaching assistantships, and other means is limited. The Flinn Foundation (Arizona) supports programs with $100,000 a year for five years to develop interdisciplinary approaches to science, such as programs for mathematicians to apply chaos theory to cardiovascular physiology. Every single program supported by the foundation chose to spend the bulk of their award on graduate students working in interdisciplinary projects. We ought to provide independence to graduate students to choose such projects by providing direct support to a much greater extent than currently done.

Floyd Bloom chaired a report for the Institute of Medicine a number of years ago, arguing that there ought to be increased support in training grants and in other kinds of fellowship support for graduate students. The report was pilloried for those recommendations. Within the last month Allen Kraut wrote a compelling editorial in *Science* reiterating our responsibility for training the next generation of scientists. This means financial support so that they can pick the project they want to work on and so that they can work at the interdisciplinary interface.

Clinical research, the capacity to apply growth substances or the products of gene research to patients, continues to be promising and exciting. But these activities require individuals engaged in research that involves patients, and in this regard we have a serious national problem. The number of M.D. investigators interested in clinical research involving hands-on involvement with the patients continues to fall. There is evidence that student debt, the way study sections are organized in the NIH, and a variety of other forces act against such career choices.

Two important responses are needed. First, we need far better training for M.D. clinical investigators. At the present time, with the exception of a program being innovated at the NIH, clinical investigators are trained in clinical programs often isolated among practitioners. We need a finite number of major centers in the United States that have a critical mass of faculty and a relevant and comprehensive curriculum within a very high quality clinical research center. By organizing such an activity on a national basis, with 8 or 10 or 12 such centers around the United States, we would: (1) create laboratories in which students could learn to do science; and (2) we would meet many of our goals for minorities and women in the field. Of course, it is much easier to create a program in which there are 100 slots nationally and distribute them by gender and to underrepresented minorities, than to create a five fellow training program. John Gallen is working on the needed approach at the NIH. It should be broadened throughout the United States.

There will also be an increasing role for Ph.D.s in clinical research. While much of the clinical research activities relevant to this discussion (such as gene therapy) require physicians working in the interest of patients, a great deal of science can be done in parallel by a Ph.D. There are now a number of programs around the country to prepare individuals for such a role. It should be noted that the M.D.-Ph.D. program is a great program, but the overwhelming majority of resulting M.D.-Ph.D.s do laboratory research, not clinical research involving people.

There are real opportunities over the next several years to obtain all payers' support of research. In contrast to the original Hatfield-Hartman proposal, I recommend that it be by an assessment to fund *clinical research.* I would argue that a 1% assessment on premiums applied to clinical research and awarded by NIH funding mechanisms makes sense. If significant amounts of money come from the health care system to fund fundamental and clinical research, it is likely that the direct Congressional appropriation for the NIH will come under pressure. There will be tendencies, regardless of any notion of maintenance of effort, to reduce direct tax support of biomedical research and to rely on the clinical care assessment. I think that's dangerous.

Many of those who deliver health care and the corporate purchasers of care argue that they already pay taxes for research and do not want to be taxed twice. We can logically argue that tax money ought to be used to support fundamental science. On the other hand, clinical research involves activities that benefit the patients, the health care providers, and the corporations that buy care. Such research is understandable, and it is proximate.

There is an additional enticement. There is a great deal of concern in the managed care industry regarding referrals -- specifically, what clinical research and which patients. They would be interested in participating if, in planning an all payers' contribution to clinical research, there was an understanding about the range of referrals and a requirement that approval of appropriate trials be based on previous peer review. With cost the big issue in managed care, today, if policy can create a level playing field with everybody paying the 1%, there may be real willingness to participate.

What does that do for basic science? At present, 30% of the NIH budget is expended on projects that fall into the category of clinical research. If, for the sake of argument, the all payers' contribution could pay for that portion of the current NIH budget which is characterized as clinical research and if the direct appropriation to the NIH is maintained by the maintenance effort, then the result would be a 50% increase in the amount of money available for fundamental science.

In summary, then, I've tried to emphasize a need for graduate education, not only to move in the direction of multidisciplinary and interdisciplinary science, but increasingly to involve the physical, mathematical, behavioral and social sciences. I believe that individual fellowships and training grant support for graduate education provides an opportunity not only for the student to have the kind of flexible broad-based experience required for a successful career, but also an opportunity for faculty and institutions to accomplish interdisciplinarity through the independently-funded graduate student. There is need for improving the quality of training of clinical investigators and potentially expanding role for Ph.D. scientists in clinical investigation. We should support an all payers' contribution to research out of the health care premium to support clinical research. At the same time, a maintenance of effort provision for the NIH should be part of any such arrangement.

4 Re-engineering the Ph.D. Octopus

Donald Kennedy

I will begin by setting out my own interest in all this. I think higher education is America's highest achievement. It has been responsible for world scientific leadership for this country, for much of our economic vigor, and for a fair share of our national cultural endowment. For half a century it has been a magnet for bright students from all over the world. Not only has that yielded a balance of payment significantly better than that of almost any other American industry, it has left us a rich and often unrecognized legacy of good will all over the world. So we are talking about a national treasure.

Now, one of the features of post-secondary education in the United States is its extraordinary pluralism and diversity. Like a complex organism, it has many specialized parts, with its own particular adaptive significance. Also like the organism, the academy has a germ line, a fraction of the whole that is critical because it is the generational link through which the basic properties of the whole thing are transmitted to the rest. That vital role in the academy is played by a relatively small number of research universities -- of which the University of California has several entries, in the extraordinary contribution it makes to doctoral training in the United States. These new Ph.D.s who spread out to the rest of the enterprise carry with them the culture they receive in the places in which they are trained. It is a culture that places high value on the production of new knowledge -- in environments where a great deal of flexibility and individual choice

---

Donald Kennedy is President Emeritus and Bing Professor of Environmental Science at Stanford University. This article is based on remarks delivered at the October 2, 1996 symposium, "Meeting the Challenge: Graduate Education in the Biological Sciences in the 21st Century" held at the University of California, San Francisco.

of direction is the rule, and in which discovery is the linch-pin. Yet the vast majority of the young men and women so trained will go to work in institutions, both academic and commercial ones, that have an entirely different mission and an entirely different set of values. That, I want to argue, is the heart of the problem we are talking about here. It is not just a job market problem, although I will spend some time talking about the 'academic job market.' The central problem is one of disparity between expectations and hopes on the one hand, and limited reality on the other. My title, "Re-engineering the Ph.D. Octopus," is taken from a phrase coined by William James back near the turn of the century shortly after the idea of research training for the doctorate, imported from Germany and grafted onto the English colleges of North America, became the primary credential for the academic profession. Even then it was plain that what we were doing was training people to do research, and then turning them out to teach. James noted that disparity and complained about it, as did many others.

Much of today's problem, and much of the disparity I spoke of between expectation and reality, is the product of an academic marketplace that is tight to the point of being congealed. We need to understand that if we are to make any headway on the remaining part of the problem. I'd like to spend some time here analyzing the history and status of that market.

Richard Atkinson's article (Chapter 1) deals with an analysis of supply and demand. I won't repeat some of the valuable points he makes in that paper, nor will I argue with him about his 1998 projections -- because I think the way you look at this market depends very much on the size of the envelope. Is it covering the academic marketplace only, or does the analysis include other non-academic alternatives that are available to Ph.D. students?

I want to begin by looking at the academic marketplace in the most restricted sense. It has three elements: retirement rates of faculty; rates of institutional expansion or contraction; and rates of Ph.D. production. It is a straightforward problem of labor supply and demand, and that is the way you have to look at it.

With respect to the first element, the vital question is when, if ever, current faculty members will end their service by retiring. Some of us are growing old ungracefully together. We are a cohort, and predictions with respect to our retirement behavior have been in for a series of nasty surprises. The first came when, as the result of the 1986 amendments to the Age Discrimination and Employment Act, mandatory retirement was abolished for members of colleges and university faculties effective 1994. There were a series of struggles between organized academia and the Congress about whether the tenure system justified a kind of blanket exemption for colleges and universities from the uncapping of mandatory retirement. It is easy to state the outcome; we lost. When I say "we", I mean the people who were arguing on behalf of the legal or fiduciary entity of the university; plainly not all faculty felt that way. Indeed, the older I get, the more I find my own viewpoints on that matter changing.

Between 1982 and the final uncapping, universities had a glimpse of the behavior of faculty members approaching retirement. Predictions were made about what was going to happen after 1994, based on what faculty members did during that interval. What actually happened depended on the type of institution in which the faculty member was working. I must take a moment here to explain to those associated with the University of California system that the rest of the world is very much less fortunate than you are: you have a defined benefit retirement plan. I envy you. We at Stanford, along with faculty members in most of the rest of the United States, have a defined contribution retirement plan -- thus at the tender mercies of TIAA/CREF until recently liberated to join Fidelity, Prudential, and other purveyors of investment vehicles. In such plans, how much we get when we quit depends on how much we and our institutions have put in. In such a program, the prospective retiree faces a complicated decision. There is a triple advantage to staying employed. First there is that extra year of salary, along with the thrill of gainful employment. The second thing you get is another year of the institutional benefit contribution. The third thing you get is that there will be one less year on that actuarial table; it'll shorten the number of years over which they are going to pay the value of your annuity.

Now during the window period, several economists looked at the prospects for faculty retirement. They were very smart people, and they went about it in the right way. But the projections turned out to be off the mark when faculty members were asked *what they intended to do*. If you ask faculty members what they intend to do when they reach 70, they don't spend a lot of time thinking about your question. They start to think harder when it is upon them; then they sharpen their pencils and perform the calculations.

Second, between the late '80's (when these projections were developed) and the early and mid-'90's there was an extraordinary constriction of economic resources available to universities. Nearly all of them were engaged in budget downsizing, like other large institutions in the society. As a consequence, expansion stopped, leaving substantially less opportunity for new appointment. The economic signals were also very discouraging to anyone considering retirement. It is much easier to launch yourself into retirement in a good economy than into a bad one.

Thus at present what we actually face is an aging faculty, fewer academic openings and the likely prospect that things will get worse from the point of view of job seekers, rather than better. That concludes a brief sketch on the demand side. I'll add just one new statistic. A highly meaningful number is how many faculty members at major research universities who reached 70 during the calendar year 1995 decided to retire, and how many decided to stay on? There are no published data yet but I have the numbers from the University of Michigan, the University of Washington, and from Stanford. In those places, over half the people who reached 70 during 1995 are staying on. That is not a hopeful signal.

Now I want to turn to the supply side, and examine the recent history of doctoral work in American universities. Before doing so, I would emphasize what a deep and often frustrating commitment the graduate experience is. It depends heavily on the good will and consciensciousness of the single mentor. It requires total immersion in a demanding scholarly discipline, often accompanied by the distraction of serving a research assistantship -- which under the best circumstances may amplify and support thesis research, but more often diverts attention from it -- or a teaching assistantship, in which the student is responsible for undergraduate instruction with varying degrees of help or guidance. It takes a long time. Even in the Golden Age of the late 1960's at the zenith of fellowship and traineeship support for graduate students in our field, the average was five years or a little more. Now it is up around seven. And the chances of failure are dauntingly high. Nationally, only about a quarter of all students who embark on the Ph.D. actually finish one. That is a terrible survival rate by any standard -- unless, of course, you are a marine fish.

In the immediate post-war years, when my own generation of scholars was in training, the horizon was limitless. The GI's were coming back; universities were expanding. Soon they were to expand further to enroll the children of those returning veterans. Federal funds were flowing in increasing amounts to support research. There was no shortage of academic employment opportunity. Today's graduate students can't imagine what it was like for us; conversely, it is a little hard for some of us to be as understanding as we ought to be about how things are for the students we are training today. That makes for a cohort effect that may have a bad influence on policy.

During the Golden Age, doctorate production in the United States increased dramatically. That growth continued until the early '70's and there was a drop-off in most fields beginning about '73. The life sciences survived that contraction much better than others, although even there the growth rate did flatten out.

If you look at the number of *programs*, instead of the total number of students, growth continued to increase even when the total number of students was decreasing. That seems at first glance a little bizarre. What was happening was that secondary institutions, still in the thrall of the Golden Age, added doctoral programs -- which are a significant source of institutional prestige. The fraction of doctorates awarded in the top-rated departments (those in the top 20 in the ACE surveys) decreased significantly.

Two interesting questions arise with respect to this period. One has to do with the paradox of decreasing production despite program expansion. How did it happen that the number of doctorates fell off? I don't think the answer is completely clear on that one. But the time required to complete the degree did rise significantly; so did the failure rates, so that is certainly part of the story. Another question has to do with declining application rates, especially for the most highly ranked programs. One possible explanation is that the loss of support for graduate study restricted the supply. You will recall that the NSF and NIH competitive national fellowships dropped. The NDEA was lost; so, on the private side, were the Woodrow fellowships. So there was some real constriction during the early '70's in that side of the support picture.

A second possibility, consistent with the rational economist's view of the world, is that entering students accurately perceived the coming downturn and selected other career avenues instead. That is plausible, but the evidence is too sparse to support it very strongly. In any event, what happened is that the number of Ph.D.s stopped falling in the mid-'80's and began a slight recovery in about 1986. Between 1986 and 1992, the number of doctorates granted increased at about 3% per year -- significantly faster in the public universities than in the private. And in 1993, Ph.D. production reached an all time high of just under 40,000 per year in all fields.

Faculty members in the research universities wanted graduate students because they are necessary to support research. As the availability of fellowship funds became limited, an increasing proportion of graduate students were supported by research and teaching assistantships. The result was a lengthening of the time to complete the degree from a national median of 5.3 years in 1968 to 7.1 years in 1993. And as the number of American students interested in doctoral study declined, the gap was filled, especially in the sciences, by graduate students from abroad.

Meanwhile, the system of academic training somehow preserved the illusion that everything was all right -- that more Ph.D.s would eventually be needed, and that if the academy could only wait things out, there would be a return to normalcy. Such views were supported by projections made by Eric Bloch as Director of the National Science Foundation who forecast big shortages in the '90's and beyond. Those studies, as you know, became the subject of bitter critical attack, and in tense Congressional hearings the numbers were challenged sharply. The depth of anger amongst young people that has followed these failed projections is really impressive. There is an electronic Young Scientists Network; to listen in on it is to audit a course of bitter discontent from cyberspace. Much of it is focused on "the myth", their term for the employment future portrayed in that NSF projection.

More thoughtful forecasts produced somewhat more optimism than was justified, but for reasons that in retrospect are entirely understandable. William Bowen and colleagues made some projections at the end of the 1980's based on anticipated retirements and concluded that there would be an excess demand over supply. Bowen believes that the main cause of the failure of demand was that new economic constraint inhibited hiring in the universities. That is certainly important, but it is also clear that the failure of retirement is a contributing factor: as I have said, retirement data now look quite alarming.

The natural consequences of all this -- the longer time to obtain the degree, the increasing residence of new doctorates in the 'holding tank' of post-doctoral fellowships -- have meant that the average age of first appointment to the faculty has risen by 2 to 3 years. Because of that and delayed retirement, the average age of the faculty has moved up--at Stanford, for example, the average age of the faculty has moved up by over six years in two decades. That is a big change, in more than one way. If you are doing the university's budget and you ask what happens if the faculty age distribution is moved six years to the right, the answer is that it costs you about $30 million. So age carries its own economic penalty. It also produces some educational and research productivity impacts that we just don't understand very well. Some claim that older faculty have more difficulty relating to students and are poorer teachers. I am skeptical about that. Others assert that older faculty decline in research productivity. The answer to that is that it apparently depends very much on the field. The data for mathematics look very different from the data in some of the social science fields, for example.

So I don't know where that takes us. I do know one thing: the training sites with the highest prestige and the highest self-confidence invariably argue that it is other, less excellent institutions that should cut back production by limiting admissions. The difficulty, of course, is that in this universe, no institution can possibly admit that it is not excellent. In the Lake Woebegon of academia, everyone is above average. A result of that shared conviction is not only a recipe for glut, it is an invitation to the kind of bitter disappointment that is broadcast over the Young Scientist's Network.

What are the consequences for our academic venture? One is a new academic class in research: the quasi-permanent research scholar. These have become increasingly significant parts of the research enterprise, even in the best places. In addition to some real intellectual enrichment, that has led to an array of problems. Scholars in this category often view themselves as just as able and experienced as tenured members of the senior faculty; yet they lack recognition and perquisites. It is not surprising that they frequently feel unfulfilled and dissatisfied -- nor that they are a significant source of grievances about academic credit, status, and intellectual property. There has also been a dramatic growth in the size of the nation's 'para-faculty'. If you look at what has happened to student/faculty ratios in the nation's colleges and universities over the past several decades, it hasn't gotten any worse, but the proportion of tenure line faculty has dropped significantly, while the proportion of what are unkindly called 'gypsy scholars' has increased. In the Northwest, the option of taking, seriatim, temporary appointments in one institution after another is called the "I-5 Route": 'Maybe this year Fresno State, maybe next year something at UC Riverside, perhaps then Portland State.' The corridor carries its own imagery of transience.

Finally, one can't talk about the production of doctoral students and the changing availability of academic opportunity without some frank discussions of the close relationship between the coupling between Ph.D. study and the support of faculty research. Graduate students, as is suggested by their frequent efforts to organize even in the best places, are seeing themselves increasingly as labor. I am afraid they are seen in the same way by some faculty. That is producing not only substantial discontent, but a marked effect on the duration of graduate study.

Not all worthwhile careers nor all serious researchers, particularly in biomedicine, are in academic settings -- although I have focused on those in this narrative. Obviously, there are attractive alternative careers in government, in industry, in the non-profit sector -- especially for biologists. But there is still a problem: it is that although the market for these newly-trained research scholars is broader than academia, that is not *their* perception of it. They have been trained in ways that make life in a research university -- a life very much like the one their major professor led – seem a life that is uniquely acceptable. To fail to get one of the prestige jobs in a research university is not only negative for the student, it doesn't help the institutional prestige of the student's major professor. As a result, there is strong pressure to clone oneself, to train (some would even say to create) a successor who will practice the same disciplinary craft in a similar institution.

So I return to a question that has absorbed us for much of this morning -- are we training too many Ph.D.s? I don't think there is a normative standard for the right number of Ph.D.s -- but it *is* plain that we are creating false expectations, by preparing more Ph.D.s *than there are jobs that appear suitable to them.* That ought to be enough to convince us that there is a serious market disparity. But we are producing too many Ph.D.s in the sense, and only in the sense, that there are more of them than there are of the jobs they are taught to want.

To return, finally, to the metaphor of the octopus: we do need to re-engineer it. But I don't think we need to take dramatic steps to alter supply and demand ratios. I think to rush into yet another series of oscillations would be risky; we have made that mistake before. Let's think instead about the kinds of changes that are needed in the way we train graduate students, and in the way we shape their expectations about what their career possibilities are.

For example, it is remarkable that we tell our graduate students almost nothing about the personal and professional challenges they'll meet in their careers, academic or otherwise. Schools of law, medicine and business now regularly offer courses in professional responsibility or the ethical challenges that they will meet in the practice of their profession. The only set of pre-professional doctoral students for whom we plan no such preparation is the one which we hope will replace us. That strikes me as a compelling irony.

Too little attention is given to the enhancement of teaching skills, or to the importance of teaching as part of academic duty. Many places are now doing a little better at that, urged by what have become systematic complaints on the part of the undergraduates about their TA's and on the part of parents about their children's education. But, at the same time the Center for Teaching and Learning is offering videotaped help on conducting calculus sections for graduate students, the major professor of those same students may be indicating in countless little ways that such activities are wasteful diversions of intellectual effort from the doctoral thesis.

The mentors from whom graduate students gain nearly all their pre-service guidance have little or no experience with the kinds of careers in which many of their charges will be engaged. Their own histories are primarily in research universities. Their ideas of what's important are formed in that culture; they're often disinterested in the activities and aspirations of their students that don't relate to those values. Discussions of pedagogy, or relationships with peers, or how you get along with colleagues later on, or how to manage a group are rare in the environments in which most graduate training takes place. That is a point Bruce Alberts (Chapter 2) already made, and I think it is worth repeating.

Finally, I close on a note of hope. I think things are changing; I perceive a new attitude in the young researchers that are emerging now. A couple of years ago when the Presidential Young Investigators, surely an elite younger generation research cadre, was asked by the National Science Board, as it is each year, to create a subject for their own symposium, they selected teaching and reform in the teaching of undergraduate science. I hope I don't need to tell you, since you're here at UCSF, that your colleague Bruce Alberts has made an extraordinary difference at the National Academy of Sciences in the attitudes of the senior scientific community. He stimulated us to stage a convocation at the National Academy in April of 1995 on undergraduate science education, and he has reorganized and strengthened the Center for Science, Math and Engineering Education at the Academy. Impressive scientific volunteers have enrolled in this effort, and that is a hopeful sign.

The opening of opportunity in non-academic fields, particularly in the sciences, is also changing attitudes and processes in university departments. It is no longer considered unusual that a new Ph.D. in biology will go off to a biotech startup, or that a chemist would decide to teach for awhile in an interesting secondary school before doing something else. These trends may eventually force faculty and departments to reconsider whether it is best to continue the present emphasis on strictly research training, or broaden it to include a more comprehensive kind of professional preparation. The attention given to this matter by the Pew

Foundation through the Preparing Future Faculty and by the nation's leading scientific academy gives me real hope that significant change is in prospect. It had better be. The present disparity between expectation and opportunity is troubling, and it is producing a morale crisis. If all of us training young researchers could broaden that training to include a respect for, and an understanding of, other possibilities, that would help. That should not be hard; it only requires the student's aspirations first, instead of one's own image of what an appropriate career is.

5 Meeting Workforce Needs

Edward Penhoet

Last semester I taught a graduate seminar at Berkeley reviewing career opportunities in biotechnology. I was frankly amazed by the poor understanding among biology students of career options available to them outside of the traditional university environment. I will attempt to shed some light on this issue by comparing the business and university environments and presenting a number of career opportunities for scientists in biotechnology.

University and Industry Now Work Together

Many industries -- high tech industries in particular -- have become increasingly dependent upon value-creation through knowledge. They therefore have a strong business need to work with the knowledge generators in universities. Universities at the same time are cooperating to transfer knowledge and technology at an unprecedented scale, and are increasingly seeking project funding support from corporations to replace dwindling public funds.

In the last decade, the businesses of America have become much more university-like than they ever have been in the past. At the same time, universities have become more sensitive to their interfaces with the business community, and many business principles have been adopted by universities. As a result, the choice of academia or industry is no longer a binary one, as it was when I began my career in science in the 1960s.

Edward Penhoet is the President and CEO of Chiron Corporation. This article is based on remarks delivered at the October 2, 1996 symposium, "Meeting the Challenge: Graduate Education in the Biological Sciences in the 21st Century" held at the University of California, San Francisco.

"The End of the Job"

An important change in business practice was recently described in a *Fortune* magazine article called "The End of the Job."[1] A traditional job used to mean that one had a traditional boss. In this context, individual employees were expected to show up each morning and find out from their bosses what they should be doing that particular day.

In contrast, today's progressive American businesses are populated by collections of professionals who take personal responsibility for achievement of the goals and objectives of the organization and who work with substantially less supervision than was the case historically. This way of working requires a strong vision and clear articulation of goals. Companies organized in this fashion typically expect to benefit in direct relationship to the contribution they make to society.

Obviously, companies expect to be profitable, but are approaching business from a societal contribution point-of-view rather than a pure financial management point-of-view. This approach is one that people with academic backgrounds can readily appreciate.

If academically oriented individuals can find meaningful and rewarding employment in industry, we are doing a disservice to them by not familiarizing them with what American businesses are like today: what career opportunities there are; how skills and careers can relate; and how the corporate environment functions.

"Whatever You are, be a Good One"

Business and science fields are clearly moving ahead very quickly. As a result, most people will have three or four different careers in a lifetime. Although specialized training is very valuable

[1] *Fortune*; September 19, 1994: author: William Bridges

for the experience it provides to young people seeking an entry level position, I believe that a general education with an emphasis on fundamentals is more useful in the long run. I ran across a fortune cookie once many years ago when I was a full-time faculty member. The fortune inside said, "Whatever You are, be a Good One." It has been the most useful advice I have ever been able to pass on to students. It is such a simple concept and so valuable. If we are going to change careers three to four times in a lifetime, the opportunities available at the next step almost always will be more dependent upon the quality of the job that was performed previously, regardless of what it was, than the direct connection of that last role to the next. Whatever specifics we teach in graduate education, we should continue to give young people the opportunity to experience doing something really well.

Biotechnology Opportunities

There are numerous career opportunities for scientists in the field of biotechnology. If one looks at the typical development cycle of products in biotechnology, it involves many different disciplines ranging from process development, engineering, manufacturing and regulatory affairs, to communications/ marketing, legal and investment banking.

It starts with *research.* Research in the biotech corporate environment does not differ fundamentally from research in the university environment today. There is a tremendous amount of cooperation across interfaces of research organizations in the academic world and in the business world. What differs, however, is the result of research. Publication is the end of the line for the typical academic researcher, but companies actually have to apply these results to create products useful to society. To do that, we in the healthcare industry have to develop processes that will allow the products to be manufactured, file an IND (Investigational New Drug application) with the federal Food and Drug Administration (FDA) so that clinical trials can be conducted to assess product safety and efficacy, and write comprehensive documents such as product license applications (PLA) and establishment license applications (ELA) to seek resulting approval for commercialization. In addition, we have to manufacture, comply with quality assurance and control processes, and create an organization to sell and market these products.

In most organizations I know, people with strong backgrounds in science can make valuable contributions throughout the entire product development and commercialization process. It was the lack of understanding about these opportunities that I found most surprising in my course last spring at Berkeley. I am a faculty member of UC Berkeley, and I spend a lot of time there. Yet, the students in my own department who have heard me talk still had no real concept of the opportunities in industry that were available to them with their training.

The following are a few examples of areas in our business of biotechnology where a science education plays an important role in a business function. They are illustrated in Figure 1.

- *Pharmacology and toxicology* is the function that evaluates efficacy and dosing for new treatments or preventative products in animal models. This group is responsible for studying and evaluating the subjects' reactions to compounds to assure the products' safety in diagnosis, treatment or prevention of a given disease. For this function, a strong background in human and animal model science research and medicine is necessary. A typical candidate might have an undergraduate degree, a Ph.D., M.D., or pharmacology degree.
- *Process development* is a discipline that requires substantial knowledge of protein purification, stabilization, formulation, and scale-up. This position interacts frequently with engineering; therefore, an engineering degree is usually required for this job.

Figure 1

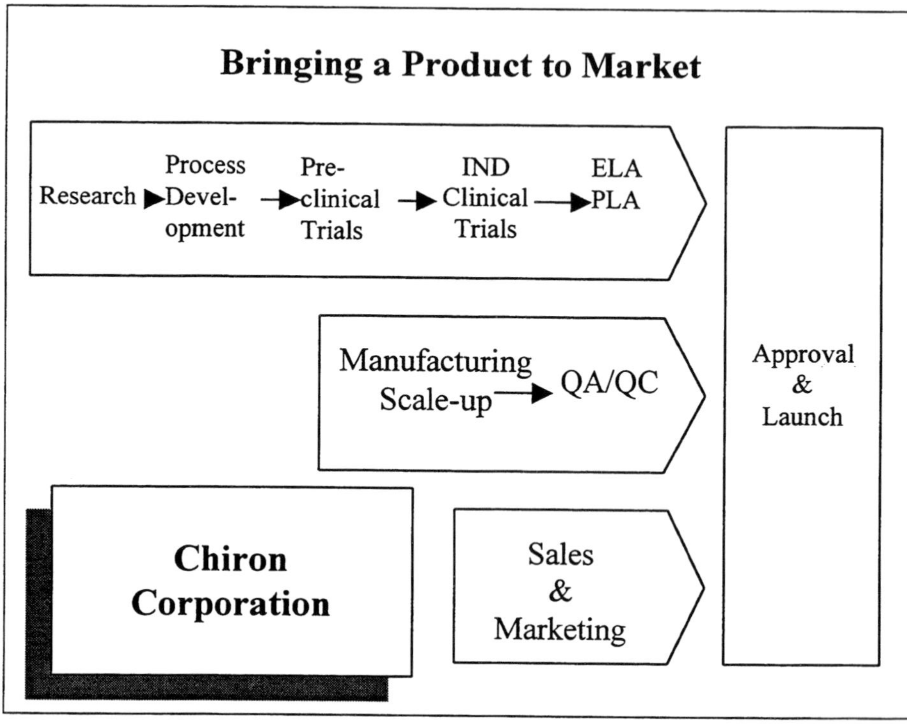

- *Clinical research* is a major focus within the biotechnology industry. Ultimately, clinical research must be supervised by a physician, but having a good scientific background is a benefit for supporting jobs. There is nothing mysterious about clinical research. It involves doing a good series of experiments on human beings to test a hypothesis. Of course, if one is experimenting on humans, the subjects tend to be a little more "touchy" than mice. Nevertheless, it is essential that human clinical trials follow the same strict standards of the scientific method as in any other type of experiment.
- *Regulatory and Quality Affairs* is the area of biotechnology which involves interacting with the FDA on behalf of the

company. The best way to deal with the FDA is with good science, not through lawyers (although, occasionally, one does need a good lawyer). An attorney used to be head of the FDA, but for the most part, people at lower levels in the FDA are solid scientists. And if one wants to go to the FDA with convincing arguments, it takes a scientist to make those arguments. The head of our regulatory and quality group at Chiron is a woman who has a Ph.D. Bernardita Mendez got her degree with John Baxter at UCSF and did a post-doctoral with Dan Koshland at UC Berkeley. She is very good at her job primarily for two reasons: she has outstanding fundamental training in science and she is a marvelous people person. She really knows how to motivate people to work hard with a purpose in her organization.

There are many career opportunities that science students don't ordinarily think about. *Communications and Marketing* are among them. It shouldn't be a surprise that if one is trying to communicate scientific business issues or if one is trying to market products that are fundamentally scientific in their application, then a person with training in science could be quite useful in that regard. Some background in science is almost mandatory for communicating what is going on in a business like ours, and particularly for marketing products to technical customers. Because our population is becoming increasingly literate about scientific issues, a science background in these fields will become more important in the future than it has been in the past.

The whole debate we have had about healthcare costs in the United States in the last few years has been more about value than any other subject. There is not a lot of debate when one talks to people about whether they want to spend money for healthcare -- the debate is about whether they get good value. Consumers do not want to waste money. But the value equation demands that every supplier demonstrate value, and, in the case of biotechnology, the value is fundamentally scientific in nature.

- *Lawyers* with scientific backgrounds also make significant contributions to biotechnology, particularly in the intellectual property area. Patents on products and their applications help biotechnology companies reap the benefits of their investments in product research. It is not necessary to have a Ph.D., but many of the best patent attorneys today do have Ph.D. degrees to complement their legal training.
- *Finance* is another field that is important to the biotechnology industry. Biotechnology has required enormous amounts of capital, and it has come from venture capitalists and public investors through investment banking firms. These groups have had to build a strong internal science capability in order to help them evaluate whether a technology or a company is worth an investment or not. In the early days, investors made their investment decisions based on the people involved in companies, but increasingly, all of the analysts -- the people who evaluate investments -- have a background in science. In fact, a very large number of physicians are now employed in the financial community.
- *Business development, society and government* roles are also important in the field of biotechnology. Articulating to society the value of the whole medical enterprise is essential in preparing society for new technical developments, such as genetic testing and gene therapy. My own company, Chiron Corporation, employs scientists in the specializations and many other positions. Indeed, if one examines the way Chiron is structured (see Figure 2), it is apparent that there is no department or position in the company where a graduate or undergraduate degree in a scientific discipline would not be advantageous.

Moving Forward

In conclusion, I believe that there is a great future for biotechnology as a field. This future will provide many job

Figure 2

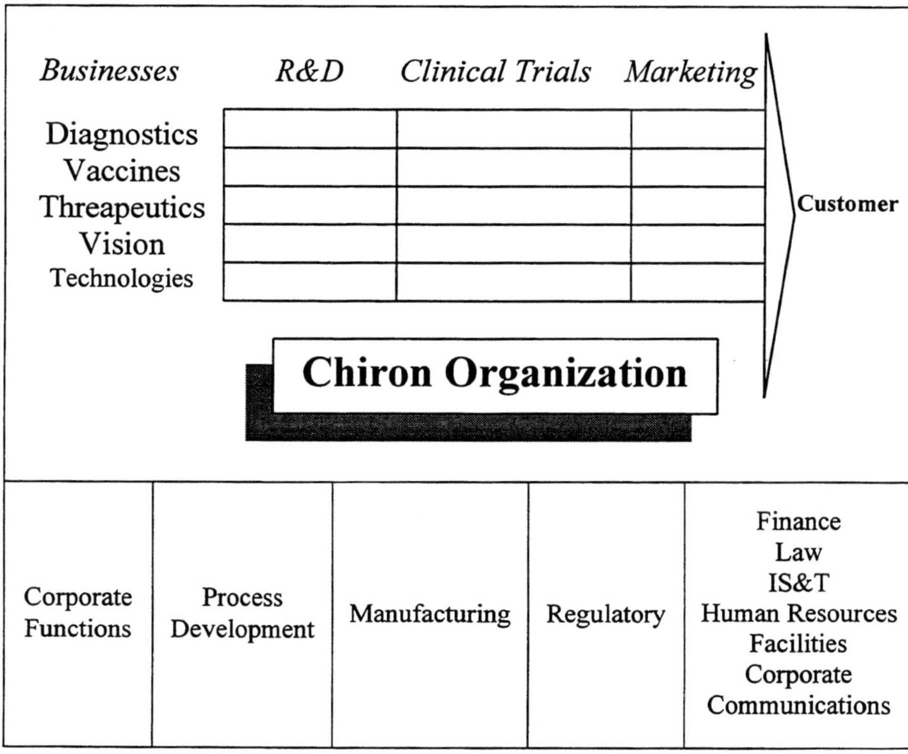

opportunities for people who understand science and its applications in healthcare. Let me return to the fortune cookie message. "Whatever you are, be a good one." I have had the opportunity to view both sides of this world -- academia and business. Although most of the details of what people my age learned as students in the 60's would have little utility today, the fundamental principles of science remain as personal cornerstones. Knowing how to think about science, how to do science, how to use the laws of physics and chemistry, and knowledge of biology, will allow people to continue to contribute to society in many different ways -- our challenge is to be sure that whatever they are, they'll be good ones.

Panel Discussion

Bruce Alberts, Donald Kennedy, C. Judson King, Edward Penhoet, Kenneth Shine

Bruce Alberts: I would like to ask Ed Penhoet if he has taken some of the information he learned from students in his course at Berkeley and made it available on the web? They are the kinds of mysterious secrets that students need to have made available to them.

Edward Penhoet: Actually, I had the students in the course write papers on each of the subjects. We went through the whole list of subjects, and I believe that we are going to do something similar at UCSF in the Spring. Out of the combination of these two sets of papers, we will either be writing a monograph or putting it on the web, or perhaps both. It will in some way be readily available. An interesting aspect of the paper assignment, apart from the students not knowing anything about the business world, was the following. In the class, there were many more students than topics. So I said that I only wanted one paper on each topic, the students would have to divide themselves up. There were 40 students and 15 topics, so this put three students together on each topic. From the three students on each topic I wanted only one paper, on which they would have to work together. This was a terrifying concept for graduate students -- that they were actually going to have a grade that was partially dependent on the performance of two other people. This pointed out something interesting: in our educational process, we are so worried about cheating and cooperating and all

This discussion is based on remarks delivered at the October 2, 1996 symposium, "Meeting the Challenge: Graduate Education in the Biological Sciences in the 21st Century" at the University of California, San Francisco.

of the rest that we have isolated these young people into their own little islands. One of the things we have to figure out is how to get students to understand that most of the world works in teams, although the academic world focuses on an individual's performance.

Kenneth Shine: I will just follow up on Ed Penhoet's comment about value. The assessment of value in health care turns out to be a very complicated, very difficult issue, yet it is one of the most important set of research issues present in the current system. Research is already being done by multidisciplinary teams. There are a few areas in which there are more people with different sorts of expertise -- economists, sociologists, as well as biologists and physiologists -- doing this kind of investigation. One of the biggest challenges in the healthcare system, for example, is to understand how to risk adjust for disease. The whole system of managed care that we are dealing with is one in which the strategy is to avoid taking care of sick people, to take care of well people only. Under these circumstances, profits will be high and costs will be low. The reason that this happens is that there is no good way, at the present time, to pay a provider for taking care of people who have very high risks of being ill. I am not discussing only clinical research. Also included is the necessity for much more sophisticated ways of measuring the quality of care that will increasingly involve a very large number of disciplines.

Thomas James, Director of Pharmaceutical Chemistry Graduate Program, UCSF: Yesterday, at UCSF, we had a site visit from an NIH training grant study section. To these people we had tried to stress the concepts provided by the Committee on Science, Engineering and Public Policy (COSEPUP) report and the fact that more of our students, in general, were going into industry in recent years than they had in the past. We indicated that we thought this was probably the trend. I was somewhat taken aback by the executive session yesterday afternoon when they took us to task for the fact that we are not preparing people for academia. I would like to ask what steps are being taken to educate these people, who are important decision making bodies, of the recent trends in graduate employment?

Kenneth Shine: First of all, in our conceptualization of this I think we have to recognize the role of the NIH in supporting science, and on the other hand the role of the NIH in supporting training. Somehow, we have gotten the two roles intimately interchanged and we don't accept the notion that the educational and training missions may ultimately be better served by looking at things in a somewhat different way, not necessarily whether or not we are funding the best specific research project. What I would point out to the people who are in this position is that a year ago, when the NIH budget was in real jeopardy, the single most pivotal events leading to significant increase in our budget were conversations between the biotechnology industry and Newt Gingrich. In these conversations, the biotechnology industry said the following: the rubric that you have been following, namely that the fundamental research goes on in the government while the developmental research goes on in industry, doesn't work for this industry. The NIH needs both fundamental research and research that yields products in order to succeed as an industry. This is not only true of biotech. If you bring a group of industrial leaders into a room, which we have done on numerous occasions, they will tell you the same thing that Donald Kennedy told you. The most important products of the university are the educated people who can work in a productive way. That kind of message takes time to get into people's consciousness. The reality is that the NIH budget has done very well for political, social, and personal reasons. But it also has done well because it relates to what goes on outside of academia.

M. R. C. Greenwood: It seems to me that some of the issues related to how the NIH evaluates things like training are part of a larger question, and that has to do with how we measure our productivity and success in science, as well as the scientific communities reluctance to address that question. We are in a cycle right now where there is reengineering of government, productivity measures, etc., being suggested. Part of what is happening is that we measure what we can measure, and one of the things the NIH tends to measure is the number of people who have been trained in their programs who subsequently get RO1 grants and go on to academic careers. We haven't been successful at suggesting to them any alternative measures of what would be an equivalent return to society, and that would then in some ways help them validate with their own legislative committees the effectiveness of their programs for the country's future. So it is this issue, of becoming what we measure, as opposed to measuring what we wish to become, that seems to me to be a crucial part of this discussion.

Donald Kennedy: It occurs to me that the COSEPUP report ought to be useful. I was going to start out this commentary by saying, "Dear Harold," because the NIH does need to hear that it is very important to measure program success in terms of the full set of outlets that constitute the mission and not just one. The COSEPUP report speaks to the breadth of training and the breadth of external sources of employment. It seems to be that the Academy and scientific community generally ought to be pushing in NIH and NSF to adopt those definitions. I think the hard problem is doing what M. R. C. Greenwood suggests -- giving them objective indicators which everybody can agree on. You have to have a tracking system that is quantitative. It seems to me that if somebody who is going to work at Chiron is publishing just as many papers in the same journals as somebody who has got an RO1, those ought to count as equal.

Edward Penhoet: I think that part of the problem, historically, was that there was a binary distribution of people in science. The people who could get the work went into academia, and the people who were failures of some sort or another were sorted into industry. So there is a long historical bias that says you must not be turning out good students if a large batch of them have gone into industry. But certainly, in today's industry, people do publish very aggressively, and that is one way to measure. There are other measures. We have to figure out some way to guide NIH in tracking the careers of these other people. I think the quality issue is very blurred at the moment. There are some very fine people on both sides.

Interestingly enough, a panel has been put together to look at the pure economics of basic biomedical research. The economists studied a number of different things, but one of them was productivity in the world of business. It turned out that the research organizations in the pharmaceutical companies most productive from a product point of view were also populated by people who were best known by their colleagues in the academic community. The standards are not really very different, but we have to somehow raise the level of awareness.

Wally Schaffer, NIH: I just want to respond to a couple of the comments that were made. The NIH does have a comprehensive evaluation project going on right now. It evaluates our training programs as have been practiced for the last twenty years and we are not limiting the study to what are typically considered to be academic success stories. We are actually going to go out and ask people how they have been involved in research and what other contributions they have made. I think that we are hearing these types of comments.

Kenneth Shine: I would like to point out that the whole realm of multi-scientific, multi-participant research, such as clinical research and health sciences research, has the same problem. That problem is that the NIH and others of us have trouble, when research is done that involves teams of people, dealing with issues of credit, productivity, and the fact that they take longer to do than other things. So it is not as if this is all just about industry versus academia; it is also about assessing the success of ventures in a variety of areas.

Roger Lovelet, Director of Science and Technology Development, Baxter Biotechnology: Baxter is a very large company and I'd like to amplify some of Ed Penhoet's comments. Ed and I actually occupied adjacent areas in Seattle one time many years ago and, like Ed, I spent 20 years before I started what I am doing now at the University of North Carolina at Chapel Hill. I still have an appointment there.

The publication issue is critical. The thing that is going to drive pharmacy and therapy committees in hospitals to approve the use of new drugs is publication in peer review journals.

Another thing that I'd like to mention is that it is going to be very difficult for a technical person to rise in a company strictly through a technical track. They will have to move to regulatory affairs, they will have to move to sales and marketing, and they will have to have a variety of experiences to be valuable to a large company. That is certainly the Baxter story.

The third point is in regards to what Ed Penhoet mentioned of the emphasis moving from research to production. About 5% of the effort at Baxter is in research. The other 95% is in manufacturing scale-up, development, and so on. Our company is increasingly looking to co-development of products. People have to be able to ask questions about products in areas that they haven't developed. That they are getting back a basic understanding of the discipline, rather than a specialized knowledge, is of critical importance.

Kenneth Shine: One of the fundamental problems that we have, though, aside from describing opportunities, is the numbers game.

We are all worried about how many PhD's should be trained and whether there will be jobs and so forth. Any time we attempt to get a handle on this, as Donald Kennedy suggested, you can make some pretty good projections in the academic environment. There is a range, but trying to get an idea as to what the real opportunities would be in industry is one of the most difficult problems we have. Do they have any ideas, whether or not it is done anonymously? Is there any mechanism by which one could begin to get a handle on what the expectations are with regard to new jobs in industry?

Edward Penhoet: It seems it would have to be in industry's best interest. I don't think it is proprietary in any way. One point to bring up is that nobody has asked me for it yet. It would take some effort to collect, but I think it is possible. I would also like to re-emphasize the point that most of the jobs I talked about today will not lead to publication. They won't lead to showing up on many of the surveys and so we would have to make sure we looked broadly to understand the impact.

Kenneth Bell, UCSF: My question is in regards to the future teaching role in universities, at both the undergraduate and the graduate level. It has to do with the fact that as teachers we are supposed to be communicators, but the model that we use for teaching really goes back probably to Aristotle -- standing in front of a class, giving lectures, and so forth. New media, for example the Internet and World Wide Web, changes, perhaps, what the future model will be. My question is whether you can tell us your opinions on how the university will look in 20 years, what will be the role of academics?

Bruce Alberts: I will address the issue of the electronic revolution. Is this an opportunity for universities? Is it a threat? Is it going to make major changes? Who will be creating the grease of that future, Microsoft and Disney, or the universities? This is a serious issue. There is a somewhat broader range of issues that are going to be considered.

Donald Kennedy: I think that anyone who predicts now the impact on instruction of the World Wide Web and the Internet runs the same risk that people who predicted in 1965 that television would completely revolutionize instruction in the universities. If you want a great prediction, walk into the largest, newest lecture halls at any university and notice a wonderful array of projection facilities and then ask your host when it was used last. There is a real tendency of these technologies to deliver a lot less than was promised.

Having said that, I don't underestimate at all the capacity of the Internet to develop both opportunities and challenges. If everybody can quickly call up Bernard Baylon's last lecture on colonial society, that is a wonderful opportunity. But the Internet is also an intellectual flea market, literally crying out for intelligent guidance. All of us will have a lot to do sorting the dross from the small fraction of worthwhile information that is out there.

My second point is the following. I have reached a point where I will no longer offer a course in which I don't ask the students to work together on projects for a significant portion of the class, to share the grade, and to deal with the free rider problem, themselves. When we begin to recognize that students have for years been teaching each other a lot more than we teach them, and get ourselves in that position of leveraging catalyzed learning, we'll continue to miss great opportunities.

Linda DeYoung, Genentech: I guess I feel there is an assumption in academia that says that you know the students are vested academics as that is what they are being trained to be. I felt, coming out of a postdoc not too long ago, that I had sort of failed by going into industry. On the counter side of that, why do we have the assumption that the scientists are going to be best at doing academic research? I thought that it was very enlightening, having come to Genentech, that there are diverse skills that scientists have who are not necessarily doing science. Some of them are very good at doing business. Some are very good at managing. Some people are good at negotiating. But they are all still scientists and they don't know about those other skills until they've gotten into industry. My question is, how can students have some insight into what their personal skills are, besides just their scientific skills, to know where they may be best placed? Are they better off in academics, or is that just a pathway they have to follow to find their place?

Bruce Alberts: When I was teaching undergraduates at Princeton, I came up with the idea that everybody had an IQ and you could measure it somehow. But then, after working with hundreds of Princeton undergraduates in my lab, I recognized that there is no such thing as intelligence. Everybody is skilled at different things and that is something that students do not understand. It is a very important message for them to get.

One of the first things we could tell our students is that there is no linear scale for abilities. Then we have to figure out ways to close in on the things that are different from the standard academic experiences. Give students a chance to do things outside the university so that they can find out for themselves what they are really good at. The first thing to tell students, and this is something that I tell my own children, is that there is no uniform thing called intelligence. People are good at different things and that is a fundamental issue.

Donald Kennedy: I think every graduate student department should have, before too long, a seminar for second or third year doctoral students, conducted by somebody who has a reasonable breadth of experience and who has wrestled with some of these challenges. It should be organized in such a way that a small number of students really interact and talk about their aspirations, talk together in such a way that they are allowed to say things that are now, in many laboratory environments, almost not permitted to be said at all. I think that once they start getting talked about and people begin to recognize that different folks in the room have different strengths and different aspirations, it becomes part of the conversation. People don't feel the way you did when you first came out.

Bruce Alberts: One thing that is relevant is that the COSEPUP report talks about exposing students to a lot of different things during their graduate careers. There is no way we can do that effectively academically if we insist on everything being one semester long. I think we should start rethinking the length of courses. There should be mini-courses, a lot of mini courses in different things. If people want to learn about computer sciences, let them start small so they don't have this big activation energy barrier to get into something new. If students could take week-long courses, for example, they could figure out the guts of something and decide whether or not they want to learn more about it. We need more kinds of courses than are presently in our graduate schools.

Edward Penhoet: Perhaps it is attitude as much as anything else. In graduate education, we are very rigid. I oftentimes joke with my friends about living in Berkeley and how much I enjoy occasionally going to Los Angeles. The reason is, in Berkeley, everything you do from the time you wake up until the end of the day has a right way and a wrong way to do it. You can only drink a certain brand of coffee. There is absolute and total rigidity. Something I like about Los Angeles is that you can do anything you want in Los Angeles.

But too often in graduate education we have this same Berkeley mentality which is binary. There is only one proper way to do something and we are very rigid in how we think about things, how to finish thoughts. Over the years, in the business that I am in, I have been visited by many people from Europe who try to understand why entrepreneurial businesses are successful in the United States and not in Europe. The biggest single factor is the fear of failure in Europe is so much higher because there is a stigma associated with it. Bankruptcy is a wonderful concept. It allows people to experiment, and if they fail they are not executed, they are only bankrupt. But you can recover. We don't have bankruptcy in the academic world. There is no easy way to fail and get up and start fresh all over again. One of the reasons why so many people are willing to try new businesses in the United States is that they are not branded forever if they go broke. Academia needs some equivalents, I think. Either have a more flexible attitude, allowing little bits and pieces here and there to be a little off, or academia is stuck on that rigid path.

Bruce Alberts: I just want to mention that graduate students in America's medical schools have been, until very recently, the most ignored players in the educational environment. They have been isolated. They have always been seen as secondary in importance to the medical students. In the last five to six years that has begun to change. The Association of American Medical Colleges has at our urging created a new organization made up of vice deans or chancellors who are responsible for the graduate education in academic health centers. They are meeting for the third time in October and one of their main priorities is this issue of students' voices. I am skeptical about what can be done on a department by department basis. Given the rigidity and inertia in the department, an opportunity to bring graduate students together from multiple places around the school and have these kinds of sessions, have this cross-fertilization from a variety of sources, is much more likely to be productive. There are departments that are terrific and could do it, but the difficulty of jumping over the local inertia is a very difficult one.

Edward Furtek, Associate Vice Chancellor-Science and Technology, UC San Diego: I want to comment on Dr. Kennedy's presentation about the generational conflict with faculty not retiring. There is another set of demographic forces which are the opposite of that -- people that are retiring. The number of people retiring is increasing so that the entitlement part of the federal budget is really squeezing out all discretionary programs. That is another factor in this generational conflict. Until the research community -- and that is broadly defined as both the private and the public sectors -- address how we get ahold of the entitlement program, we are not going to have discretionary funds, whether or not there is an all payers' premium. Entitlement is just too hard on the budget.

Donald Kennedy: There is good news and bad news. The good news is that the cohort that is retiring now is the best fixed and thinnest cohort we are going to see. Whatever problems the entitlement programs face now are minor compared to the problems they are going to face when the baby-boomers start to retire. That, I think, is going to force a re-evaluation of the entitlements picture before they get there and I hope that it does.

6 Plants That Croak: A Job Description for Genetic Engineers

David Perlman

I have been told, by people who ought to know, that graduate education in the biological sciences is a rat race, and that at the end of it most often is a choice between Academe and Industry -- although I have met more than one Ph.D. driving a cab to the airport.

In the Academy, tenure is the goal, parking is the major problem, and campus politics are the principal lunch time diet at the faculty club. Nor does commerce seem much better: will a start-up company's venture capitalists patiently support basic research as well as product goals? Will the scramble for patents inhibit free exchange on the Internet? Do you go where the money is, or where the unfettered scientific challenges lie?

I am not going to offer journalism as the noblest profession free of all such issues. Rather, I am going to discuss something of what we journalists do -- what, if anything, we can contribute to the public's understanding of science. I am also going to discuss why our state of California badly needs a generation of trained people who can more effectively help the public to understand the nature of scientific progress and to make more informed political decisions on issues that science poses for all of us.

For example, let me offer a brief comment on a currently hot political issue: illegal immigration, California's proposition 187,

---

David Perlman is the Science Editor of the San Francisco Chronicle. *This article is based on remarks delivered at the October 2, 1996 symposium, "Meeting the Challenge: Graduate Education in the Biological Sciences in the 21st Century" held at the University of California, San Francisco.*

and their relation to decision-making based on an understanding of scientific -- or at least medical -- issues.

Our governor has indicated -- with considerable public support, I am afraid -- that undocumented alien women who become pregnant should be denied prenatal medical care. A wise economic decision, after all, because the taxpayers will save money, and the maternity clinics in our overcrowded public hospitals will have more room to care for the more worthy of the poor among us.

But how many ordinary California voters are aware of the medical advances that can avert the dangers of problems like preeclampsia? How many realize that prenatal detection of birth defects can render postnatal intervention far more effective? Or that simply preventing low birth weight in infants can make it unnecessary to mobilize perinatologists and intensive care nurseries for emergencies that are more costly to the taxpayers than simple periodic visits to a primary care physician?

Recently, on KQED radio's morning talk show, Forum, I heard both pro and con advocates agree that whatever disputes they might have, removing children of illegal aliens would certainly be "cost-effective!" Neither debater, it seems, could understand the ultimate social cost of more thousands of Californians growing up virtually illiterate, most probably unhealthy, and surely unqualified for any kind of employment later on.

This is not a political paper, nor does Proposition 187 have much to do with Science with a capital S. It does have to do with *rationality*, however, and that word must always come into play when science issues -- genetic testing, RU-486, science-based stewardship of stockpiled nuclear weapons, the teaching of evolution, silicon breast implants, you can name your own issue -- arise. I want to underscore how ill-prepared most of our citizenry -- even governors -- can be to deal with complex public policy issues involving biomedical advances. These are issues where slogans so often overwhelm rationality.

The reader may be familiar with the work of Jon Miller of Northern Illinois University who studies the public's comprehension of science by surveying random population samples with more than enough care to satisfy the criteria of statistical significance. "Nine out of ten citizens lack the scientific literacy to understand and participate in the formulation of public policy on a very important segment of the national political agenda," Miller said at a recent meeting of the American Association for the Advancement of Science. He found that two-thirds of the adults in his surveys believe that lasers work by focusing sound waves. Fifty-seven percent of those surveyed believe that electrons are bigger than atoms, 75 percent think antibiotics kill viruses as well as bacteria, 67 percent believe that the earliest humans lived at the same time as dinosaurs, and nearly half of those surveyed -- 43 percent -- believe that astrology is "very or somewhat scientific." But worst of all, 54 per cent of American adults say they do *not* believe the statement that "human beings, as we know them today, evolved from earlier species of animals." (Now do you wonder why our California school districts and our state education people must still battle every year over the dirty word "evolution" in our biology textbooks?)

This problem of scientific illiteracy, of course, is by no means limited to the so-called general public. You may recall the early days of recombinant DNA research when -- with commendable concern for many of the unknowns -- researchers in molecular genetics called for care in pursuing their work. They recommended a series of restraints that included creating the RAC, the federal government's Recombinant Advisory Committee. They set the first temporary containment standards for laboratories where scientists were moving DNA from one organism to another, and they assigned one group of colleagues to develop "fail-safe" vectors -- presumably to protect us all against the emergence of home-made Andromeda strains right here on earth.

That historic meeting at Asilomar took place in February, 1975, and within a year our own State Legislature and our State Health

Department began wondering whether the force of state law shouldn't rigidly regulate scientific research in this area. I remember covering a series of joint Senate-Assembly hearings on the issue in Sacramento, and the debate was contentious indeed. As he opened the hearing, I recall, the joint committee chairman, Assemblyman Charles Warren, declared, "While I hope that no member of the legislature will rush to write legislation, if the scientists can start crossing the genes of plants and toads, and I find myself with a philodendron that croaks, I want to know a lot more about what's going on!" Just to sample the nature of the scientific testimony given at that hearing, let me quote a very distinguished scientist, then at the Sloane-Kettering Memorial Cancer Center in New York and whom charity compels me not to name, who said, "We will have to recognize that the future of the entire human race is at stake. I think the risks are great enough that we should stop the research now."

Without recalling details of the widespread uninformed hysteria that greeted field tests of genetically engineered Frostban on California potatoes and strawberries not long ago, I think I have made a reasonable case for the deplorable and often dangerous lack of public understanding of science.

But what to do about it?

Others at this meeting have brilliant proposals for improving familiarity with science and scientific methods in every American classroom from kindergarten on. Our museums -- like the Bay Area's own California Academy of Sciences, Exploratorium, Lawrence Hall of Science, and Josephine Randall Junior Museum -- can play important roles too. Television programs like Nova on PBS, or the excellent material on A&E, or the all-too-infrequent science specials on the major networks are also important, even though they primarily reach audiences that are already most receptive to science programming.

And what about journalism?

Here is an area, I believe, where far too few people who are well trained in the sciences are functioning. And by journalism I don't merely mean newspapers and television. I include all the media of public communication: There are good articles on science in general circulation magazines that range from Time and Newsweek to the Reader's Digest, Popular Mechanics and even KQED's Focus. There are explanatory, but rarely exciting, materials that accompany museum exhibits. And there are interactive presentations on CD-Rom programs that could well attract more and more kids after they finally weary of Ninja Turtles or the Mortal Kombat bestiary.

I am a science reporter, and I have to confess I have had no scientific training beyond the Mickey Mouse science survey course I was required to take as a Columbia undergraduate: a semester each of physics, chemistry, and biology, plus half a semester each of astronomy and geology.

Fortunately, however, I became immensely curious and intrigued about science years ago when I first asked a knowledgeable friend to explain just why Sputnik didn't fall down, and then had the good fortune to interview a Berkeley astronomer at Lick Observatory, who showed me where new stars are being born out of the gas and dust in the Orion nebula. Talk about inspiring! And then Charles Weaver took me to Hat Creek, and I learned that organic molecules were whirling around out there in the Milky Way.

I didn't get to biomedical matters until I was assigned to cover a medical convention and had to ask a specialist in liver disease whether it wasn't unethical to mention the name of a sick patient she had tested when -- as I thought -- she was telling me about someone named Billy Rubin! It was only a bit later that I learned how to spell deoxyribonucleic acid.

That kind of experience typifies many of my colleagues in science writing today; we got to do what we do because we were interested and tried hard to learn on the job. There *is* a better way, however, and fortunately, more and more science journalists are now entering the field after significant exposure to the sciences in their undergraduate years, and sometimes even after finishing graduate school. But even I have learned something about the concept of falsification, about statistical significance and double-blind experiments and peer review, and about which journals are appropriately refereed. And because I am paid to exercise the same instinct that all good reporters are supposed to possess, I believe I can ask some of the right questions and even smell hype when I hear it.

Hype -- overstatement, false promises, the cry of "breakthrough" -- can come from everywhere: from universities, from politicians exploiting scientists, from science-based corporations trading on NASDAQ, and even from scientists themselves.

One may remember the early days of the AIDS epidemic in America, when Gallo and Montagnier were trading insults and accusations of larceny. Let me recall one specific day -- it was April 23, 1984 -- when Gallo's team was about to publish the first report on the isolation of the AIDS virus in the journal Science. The National Cancer Institute had called a press conference for Gallo several days before publication of the article, but politically savvy media experts at the higher-level National Institutes of Health decided the story would have much more impact, and gain more credit for the Institutes, if it were called by the NIH director himself -- fear not, that was a full decade before Harold Varmus arrived! But finally the spin doctors in the Reagan Administration took over and decreed that the press conference should be held downtown in Washington instead of in Bethesda, and they appointed Margaret Heckler, then President Reagan's Secretary of Health and Human Services, to sound the good news. "The arrow...aimed and fired at the disease AIDS has hit the target," proclaimed Ms. Heckler. A vaccine to prevent the disease, she continued, would be ready for clinical trials "in about two years." And it was a good four years later that the U.S. Surgeon General, Dr. Everett C. Koop, warned the public succinctly that vaccine development can be an incredibly complex process fraught with many pitfalls -- and that an AIDS vaccine would most probably not be available before well into the coming century.

And now, today, confusion surrounds all the new AIDS drugs, and we must try to explain what protease inhibitors are when our audiences still have no awareness of the limitations of AZT. There is confusion over the Tamoxifen trials, and the cloning of BRCA-1 and what that means to women with cancer in their families. There was plenty of confusion, surely, over the DNA evidence in the O.J. Simpson trial -- and not least, within the jury! And of course the Human Genome Project is wholly arcane to most people.

And what do journalists do with the big breakthroughs?

Here's a page 1 lead of a *Washington Post* story:

> A new cancer treatment that activates the body's defenses, turning white blood cells into 'killer cells' that attack tumors, has shown highly promising results in its first experimental test in humans, the National Cancer Institute announced yesterday....'It is the most interesting and exciting biological therapy we have seen so far,' said NCI Director Dr. Vincent DeVita Jr. 'It is working against very common malignancies and showing a very powerful effect.'

That "treatment" was IL-2, and it caused a wholly unwarranted rush of patients clamoring for it in every cancer center from the NCI to UCSF. The daily press was at fault, surely -- but what was the source of its guidance?

This problem -- of over-selling disease treatment developments in the news -- is caused, at least in part, by a combination of deadline-driven journalism, corporate press-agentry, and -- all to often -- the optimism of biomedical researchers themselves. And then, of course, there is the mass confusion sown by the supermarket tabloids and the overwhelming market demand for books on health and diet and fitness. These are not contributions to the public's understanding of anything -- whether it be medicine, biomedical research or health. Nor do we need any more purveyors of such nonsense.

There is a vitally important role that young people trained in the sciences can play in the public's awareness of what science is, how the scientific world works, how true science leads to technology, and technology's dilemmas of benefits and social problems.

Most of my colleagues in journalism and I spend a lot of time interviewing scientists. Many are extremely lucid as they explain their work to us. Many are shy or diffident. Some are reluctant. And a few simply won't talk at all -- not even to explain the difficult details of a published paper that the public really needs to understand. And, unfortunately, even among those willing to explain their work and its significance, far too many simply can't seem to speak English at all.

So I have two proposals to make:

First, that in graduate science education there be a new effort to have every Ph.D. candidate either pass some kind of proficiency test in oral and written communication to a lay audience, or undergo remedial training in that kind of communication. For example, at the University of California in San Francisco many years ago the late Dr. Julius Comroe at the Cardiovascular Research Institute insisted that all the CVRI fellows take his full course in writing scientific papers. I sat in on the course, and a great deal of Comroe's effort went into making the cardiology Fellows write in just plain English -- the English as enunciated in Strunk and White and Fowler. That is the kind of training that should stand every postdoc in good stead as he or she fulfills what I believe to be part of every scientist's duty: to let the rest of the world in on what scientists do and how they do it.

My second proposal is for those graduate students who, during the course of their academic work, find their interests shifting from doing science to talking or writing about it. I meet them all the time -- young Ph.D.s and MD's, who ask me how to get into that arcane, exotic field called journalism. I discourage most of them, of course, but I tell them that there are, in fact, many opportunities in science writing -- in journalism itself, in free-lancing for magazines and book publishers, and in universities and science-based corporations where clear explanations are needed badly.

I have even sent a few likely candidates off to the University of California at Santa Cruz, which happens to have the country's sole science writing program specifically tailored for science graduates. It is not a journalism school, and its one-year program doesn't even offer a degree -- which is a real shame, I feel -- but it requires at least an undergraduate degree in science; it is highly selective, and it is far and away the most outstanding writing program in the country. Its graduates are now successfully engaged in filling the need that I have been saying here is absolutely critical as the progress of science raises more and more complex issues that only an informed citizenry can resolve.

Every field of science and technology demands some degree of public understanding -- from recombinant DNA research to cultural anthropology to high-energy physics. I hope that as we consider the challenges of biomedical education in the coming century, we won't forget the challenge of the English language as a path to reaching the far wider public that lies -- and votes -- far beyond the laboratories.

Section II

California's Future Workforce Needs in the Biological Sciences

Susan Hackwood

This section is devoted to a realistic assessment of the current and future workforce needs in the biological sciences. The articles identify the emerging diversity in career paths. They question the numbers of bachelors, masters, and Ph.D. degree graduates needed in the biological sciences. They also question the goals and impact of contemporary graduate training programs. They reflect many voices, from industry, academia, and – most important – graduate students and postdoctoral trainees.

California has long been a major innovative force in science and technology. Two of our major contributing factors to this ability to innovate are the quality of research programs provided by our academic institutions and the capacity to harness the entrepreneurial spirit inherent in our citizens. Graduate academic programs are, to a large extent, dependent on the research funding available. The expansion of industry and the level of entrepreneurial activity creates the demand for students. These two factors are related and will be more so in the future as academic programs become more closely tied to industry.

At a time when federal support for research is dwindling, there is an obvious concern for academic and research laboratory support of basic research. According to a recent analysis done by the American Association for the Advancement of Science, if the budgetary scenario currently being debated at the federal level is followed, non defense R&D would be funded at 19.1 percent less than the original FY 1995 level in FY 2002, after adjusting for expected inflation. This clearly means that there is less research

---

Susan Hackwood is Chairman of the California Council on Science and Technology. This article is based on remarks delivered at the October 2, 1996 symposium, "Meeting the Challenge: Graduate Education in the Biological Sciences in the 21st Century" held at the University of California, San Francisco.

money to support graduate students in academic institutions. Research can not go on as business as usual.

If we compare California's expenditure in R&D to the rest of the world, we can see the following:

Country	R&D Expenditure per person	% funding from industry
Japan	$920	80%
Germany	$660	70%
USA	$600	52%

The total funding for R&D in California is difficult to quantify, but is certainly less than Japan and Germany. If we are to maintain a technological leadership, perhaps we should be identifying mechanisms of supporting R&D to the level of about $1000 per person, of which 80% should come from industry.

In addition to this funding problem, in the biological sciences industries are still emerging and few are mature. The shift in importance from the *first to market* to successful *cost effective manufacturing* tilts the demand curve away from the recruiting of students trained in basic research.

Basic vs. Applied Research

When science research reaches the stage of feeding technology development well, the amount of basic research required to sustain the innovation process may not be as great as initially needed. Therefore, fewer Ph.D.s are required and more applied research conducted by masters students is needed.

There is therefore, for reasons of *changes in funding* and *changes in demand*, a shift from a focus on graduating students with basic research skills to graduating students with a broader perspective of manufacturing and production processes.

Similar shifts have occurred in other, more mature technologies such as semiconductor manufacturing and telecommunications. Industry becomes viewed as a customer and a source of funding. In terms of their relationship to industry, these disciplines all have evolved variations of the following features:

- Close ties to industry and the formation of a wide range of *academic/industry partnerships*.
- Close *feedback to academia* from the industries recruiting graduates to the nature and content of a student's education.
- A pro-active "research brokering" for *technology transfer* between academia and industry.
- Development of *non-traditional* public/private partnerships.
- Development of *"integrated programs"* where students develop cross-disciplinary skills.
- A high percentage of graduate students terminating their education *at the masters level.*

Large vs. Small Science

There has been an increasing trend over the last two decades to reward faculty for forming major *research centers*. There are some important research programs, such as the human genome project, that could not exist without the formation of a large research program. However, small, easily obtained research and development grants have spawned the innovation process and entrepreneurship that generates new industries.

It may be true in the biological sciences, as in other disciplines, that a new balance needs to be achieved between large and small science. The *individual researcher* conducting research in his or her laboratory with a Ph.D. student vs. the multimillion dollar research program that requires a full time administrator to run it. At the federal level, the pendulum is beginning to swing back in the direction of the individual researcher.

7 The Future of the California Dream: Biotechnology's Shift from Research to Manufacturing

David Gollaher

I am an historian and if you ask an historian to talk about the future of the California workforce you should remember that historians have a hard time predicting the past. What I discuss here should be taken with this in mind.

Every year, the California Healthcare Institute undertakes an analysis with KPMG Peat Marwick that characterizes the scope and scale of the health care technology industry in California. There is not currently a good single term for this industry because it encompasses basic science, medical device instrumentation development, biotechnology, and everyone from a bench scientist working at UCSF or Stanford to a pharmaceutical sales representative in the field. The article by Edward Penhoet (Chapter 5) indicates the broad spectrum of employment opportunities in the medical technology industry.

But there is a problem within the biotechnology industry. Compare it to the automobile industry. Everybody in Michigan who works on the production of a car -- whether it is making a new alloy for a piston, a car salesperson, or somebody placing the spare tire -- understands that they are part of the auto industry. But in our collective industry, people don't have that same kind of developed consciousness. To some degree, this represents a problem. It certainly represents a political problem.

David Gollaher is the CEO of the California Healthcare Institute. This article is based on remarks delivered at the October 2, 1996 symposium, "Meeting the Challenge: Graduate Education in the Biological Sciences in the 21st Century" held at the University of California, San Francisco.

There has been a lot of discussion about the changing paradigm in graduate education, and about the need to change the educational paradigm to cope with a changing marketplace. The commercial paradigm for success is also changing very rapidly, from basic R&D to applied R&D and manufacturing. The employment marketplace will look dramatically different as the biotechnology industry shifts from being predominantly research, with products in the pipeline, to one that has to manufacture its products and deliver them into a competitive marketplace.

Some perspective can be given to this concept of moving from R&D to manufacturing by a conversation I had recently with a CHI board member, Brook Byers of Kleiner Perkins Caulfield and Byers -- one of the best known venture capital firms in Silicon Valley and responsible for billions and billions of dollars of start-ups, both in software and computers, and in biotechnology and medical devices. In the course of our conversation it occurred to me that in some sense the biotechnology industry is California's other software business. But the thing that makes biotechnology different from computer software is the long lead time. From the idea or the discovery on the bench, through the FDA process of approval, to that product being available to the patients on the bedside takes much more time than the 12 to 24 month cycles in software production development. Brook and I talked about Jim Clark, the founder of Silicon Graphics, Netscape, and most recently of a new company many of you may not have heard of, Healthion. Silicon Graphics employs thousands of people, Netscape employs hundreds of people (soon to be thousands of people), and Healthion employs 80 people (including 50 engineers). The 80 at Healthion are crafting a new Internet product for employee benefits that will mediate complicated information about health care coverage between employers, patients, physicians, providers, and so forth.

Aside from a few companies like Genentech, Amgen, and Chiron, most of our biotechnology industry, right now, is in the Healthion phase -- dominated by scientists and technical people. Where it will be a few years hence is much closer to the Silicon

Graphics and Netscape phases. What will employment look like when that happens?

In California, today, biotechnology employs more people than any other technology based industry, including motion pictures and software industries, and with the exception of aerospace. Of course, aerospace jobs are in decline, whereas we are adding jobs every year in healthcare technology-based jobs.

The jobs are fairly diversified as you move from basic research, which accounts for one in six jobs, through medical devices, instruments, and diagnostics, which accounts for 28%. In this industry are also a lot of suppliers and service providers. This indicates that there is a deep infrastructure and companies have diverse needs.

California's biomedical industry is very lucrative when you have a product. The gross profit margins are great. Average sales in biotechnology are $58,000 a year and in laboratory services are $38,000. The average salary across the board is about $50,000. These are the kinds of salaries that get politicians' attention. They are also the kinds of salaries that can allow people to afford an average California home.

If you think historically about the economic structure of any industry in terms of allowing people to live the kind of life that all of us want to live, it was the aerospace industry that typified the California dream. When I was growing up in San Diego, my father was an aerospace engineer. He was able to put his kids through college and to afford a nice home. We could take vacations and live in what we now reflect upon as middle class prosperity. There is a lot of concern in California about whether that dream will wither, about where the next industries will come from that could provide the income that people will need in the future. The biomedical R&D industry is one component of an extraordinary resurgence for California's economy during the next century.

Another positive aspect of the biotechnology industry, of course, is that it is an export industry that contributes to a favorable balance of trade, even more than defense does. America enjoys world leadership in biomedical R&D. To the extent that we keep it and don't squander it, we will continue to enjoy that leadership.

With respect to jobs by business function, basic research is presently the largest employment category. This has to do with the comparative immaturity of the industry. Clearly as we move into a more manufacturing-intensive stage with production jobs coming on line, manufacturing, administration, and, of course, sales, marketing and other functions will become comparatively important. Science will continue to be the base and the seedbed for biotechnology, but as the industry matures, manufacturing and corporate administration, sales and marketing will surge as sources of jobs.

To get some sense of the promise of this industry, consider how much is in the development pipeline. Six hundred fifty-nine inventions are making their way in California firms toward the market, toward the time when they will be manufactured. Not all of them will make it, but the lion's share will. As they move through phase two and phase three clinical trials, most of them will succeed. What this means is that a vital series of products will move into the market. As this happens, companies will decide whether to build manufacturing facilities in California or elsewhere. There is fierce global competition for these jobs because everybody knows the economic benefits: high wages, comparatively clean manufacturing, high portability of product, high gross profit margins, a favorable balance of trade. Europe is involved in an intense competition to move manufacturing jobs created by U.S. biotechnology firms to Ireland, the Netherlands, Germany, and so forth. There is state by state competition as well. Texas, North Carolina's Triangle Park, and many other regions are competing aggressively with tax incentives and all kinds of inducements to lure these jobs away from California.

At present, California has about one-third of the entire U.S. biotechnology industry. In sheer number of companies, about 42% of all U.S. biotechnology jobs are in California. We also have about one in three medical device companies, including manufacturing plants of large multinational firms like Baxter, Medtronic and Abbott Labs.

Without anyone intending to build this industry here, without any central planning, California has inherited an extraordinary industry that can be the basis for a vibrant economy in the 21st century.

I'd like to conclude with an observation based on a book, *The Competitive Advantages of Nations*, that Harvard Business School Professor Michael Portor wrote. Portor asked an interesting question: "Why is it that some nations, over generations and over centuries, dominate certain industries? Why, for example, do the Swiss still lead the market in chocolate? They don't produce cocoa beans. Why does Italy dominate the market in finished leather goods, even though they don't produce hides and they import them mainly from South America?" Why, for that matter, does Hollywood still dominate the movie industry, one of the most portable kinds of industries you could imagine. The answer, Portor says, has to do with the deeper structure underneath the visible industry -- the vibrancy of competition among suppliers competing with each other to make inputs for the companies that you do see. The sophistication of local consumers, with their highly developed preferences, forces betterness and competition within the specialized local industries. You can see this if you go to Hollywood. The most important thing about the workforce in Hollywood is not that Michelle Pfieffer lives in L.A., but that there are 100,000 people there in the movie business. There are people who wait tables and act as extras. There are people who do one illustration a year -- but it is a great illustration. There is a vast cadet corps of people who produce inputs that a producer can't find anywhere else in the world. Though it is cheaper to go to Miami or to Venice, movie producers don't go because those places do not offer the same human resources.

In California biotechnology we have a kind of intellectual capital that is simply unparalleled anywhere else. If you go down the Interstate-5 corridor, what you see is the most remarkable set of inputs that a life sciences industry could have. It is the envy of the world, and it is based on the university research that has grown up around the University of California and our great private universities. It has developed an amazing array of companies, amazing in their diversity and in the way that they have leveraged with venture money the intellectual capital from the university community.

As the biotechnology industry makes the transition to manufacturing, we will have the chance to capitalize on the extraordinary investment and opportunity we have had -- and we will succeed. But it will take a political commitment on all of our parts to make sure that we don't give away our birthright.

8 Meeting the Challenge: Graduate Education in the Biological Sciences in the 21st Century

M.J. Finley Austin

My personal experience with the job market it that it is tough, very tough. It was this experience that led me to ask why getting started in a career has to be so tough and led me to take an active interest in how people are trained for science careers. After all, when I was in graduate school in the late 1980's, the NSF predicted that there was going to be a shortage of science Ph.D.s in this country by the mid 1990's. I graduated in 1989 and went into my research postdoc thinking my future in science was bright and (especially after I was awarded an NIH Individual NRSA Postdoctoral Fellowship) that I was on the road to becoming an independent investigator. Everything I heard indicated that when my postdoc was up jobs would be aplenty. When I started applying for jobs, however, I quickly found that applicants were plenty and research jobs few -- both in academia and industry. If I ever got a rejection letter it usually stated not to be disheartened because the employer had received 200 or 300 applications.

I had always been interested in policy so, in the fourth year of my postdoc, I decided to apply for several of the science policy fellowships offered by the American Association for the Advancement of Science as a back-up plan while I continued to hunt for a research job. Another fellowship is always better than

---

M.J. Finley Austin is President of the Association for Science Professionals. This article is based on remarks delivered at the October 2, 1996 symposium, "Meeting the Challenge: Graduate Education in the Biological Sciences in the 21st Century" held at the University of California, San Francisco.

unemployment, I reasoned; besides, it would give me a chance to go to Washington and fix things.

Although I soon discovered that the competition for these fellowships rivals that of getting a job, I had succeeded in getting a fellowship and failed to get a job by the end of my 4th year postdocing. So off I went to Washington -- half excited and half wondering if I hadn't just ruined my life. I love science and at the time I thought leaving the bench meant leaving science, so it was a gut-wrenching decision. What I found, however, when I got to D.C., was that I was not alone. I found a growing discontent about job availability and job quality in research. And I found it broadly, especially among other early career scientists, and not just among my fellow AAAS Fellows, but among many I met still in academia or working in local federal R&D settings.

I also found that the policy discussions around this issue primarily consisted of denial. If one would venture that there might be an employment problem in science, an easy remedy was offered: I heard that Ph.D.s only needed to seek an "alternative career" in science because there were alternative careers aplenty. And I believed this, until I started applying for these careers and received rejection letters stating the same astronomical numbers of applicants I was competing with when I was turned down for "traditional" jobs in science. In time, however, I did obtain my current position as a program officer. The value-added experience I gained as an AAAS Fellow was critical to my obtaining this job. (I urge the many biological/biomedical research professional societies that currently do not, to start offering AAAS Congressional Fellowships. This segment of the scientific community is poorly represented in this program.)

My experience in looking for a job and my love of science forced me to become active in this issue. It was not difficult for me to connect with other "young" scientists from around the country who were also concerned -- not merely about their own job futures, but about what all this would mean for science in the long-term. Thanks to Kevin Aylesworth, the founder of the Young Scientists

Network, voices were beginning to be heard, at least on the internet. The Association of Science Professionals grew as a more formal organization working to bring this voice to the policy-makers. Since we are part of the demographic group having to grapple with starting careers in the rapidly changing job market we can offer unique insights into necessary training reforms.

We believe that a critical piece, however, has been left out of discussions of how to improve the education of scientists -- postdoctoral traineeships. Very little is ever said about postdocs, very little is known about postdocs, and very little is being done to find out about postdocs. While talk of reforms is almost always framed in terms of graduate education, "traineeship" now lasts through graduate school and one or more postdocs. We would like to see more attention focused on those trained scientists who are now in "postdoc limbo", trying to start a career in a less than welcoming job market.

Before we prescribe changes for the next century, we need to have an accurate picture of the close of this century. What is known about who's getting a Ph.D. and then a job? Not enough. But the NSF 1996 Science and Engineering Indicators, which describes American science as it was in 1993, provides the best and newest data.

From this data we do know that time to degree has increased over the years (and possibly time spent in postdoctoral trainee positions) Therefore, time to first "real" job has increased. Also, graduate enrollment in the natural sciences (defined as including all physical, earth, atmospheric, oceanographic, biological, and agricultural sciences) has risen steadily from 102,969 enrollees in 1983 to an all time high of 119,769 in 1993. The 1993 enrollment figure represents a 2.4% increase over 1992, however, this increase is down slightly from the 3.6% increase seen in 1992 as compared to 1991.

Foreign citizen enrollment has also steadily increased in this category, but does not account for all of the enrollment growth.

Eighteen thousand two hundred sixty-eight foreign citizens were enrolled as graduate students in 1983 and their enrollment reached a high of 31,815 in 1992, but showed a decline to 31,273 enrollees in 1993. According to NSF, typically 30 to 35% of these students receive firm offers to remain in the U.S. NSF speculates that as emerging countries expand their capacity to educate at the doctoral level, fewer foreign students will come to the United States to be educated.

The actual number of Ph.D.s earned in the natural sciences has grown from 7,676 in 1977 to 10,354 in 1993 (about 35% increase). When examining earned doctoral degrees, the NSF report breaks down the natural sciences category a bit further, so we can look more specifically at biological and agricultural sciences which are lumped together. In 1975 4,402 Ph.D.s were awarded in this group and, although some years showed slight declines over previous years, overall there has been a continual growth in Ph.D. production in these broad fields. In 1993 6,058 Ph.D.s were awarded, a record high and almost a 38% increase since 1975. This is also a 3.25% increase over the number granted in 1992.

What happens to these Ph.D.s? For Ph.D.s in the life sciences, most now plan to go on to postdocs. According to NSF, in 1970, 39.9% of this cohort with definite plans at graduation reported planning to go to a postdoc. This proportion has risen to 70% in 1993.

We also know from the Committee on Science, Engineering and Public Policy (COSEPUP) analysis of NSF data reported in the "Reshaping Graduate Education" report that from 1977 to 1991 employment of life sciences Ph.D.s 5 to 8 years after receiving their degrees has decreased from 58.2% to 46.5% in academe, and has decreased from 9.7% to 6.7% in the federal government, while there has been an increase from 12.9% to 26.4% in employment in business and industry. Furthermore, the category of "other" employment has grown.

From this we can conclude that the job market has continually changed, and that training trends and hiring both into postdocs and into permanent jobs have continually changed during the same time period. Now we have to ask: have these changes been mutually supportive or diametrically opposed or somewhere in the middle? Do we need only to adjust graduate training or do we need to adjust the entire training structure? I would like to propose that we need to rethink training in general. Is it truly serving the purpose of training or is it, as some maintain, a way to maintain a pool of low paid workers?

In order to enact useful reforms we need to know more about postdocs: How many are there? Where are they? What is the quality of their training? What do they get paid and by whom? Do they have benefits? Who are they accountable to and, more importantly, who is accountable to them? What is this training preparing them to do? What are the actual jobs these individuals are finding? What is the quality of these jobs? And, why are there industry employers saying they cannot find employees while there are postdocs and graduates saying they would like to work in industry but cannot find jobs there?

The COSEPUP report called for better data collection about training and employment and for more timely dissemination of this data. Students need this data so that they can make better training choices. Several of the physical sciences professional societies have a history of good field tracking and provide their constituents with sound timely data. The Commission on Professionals in Science and Technology is working to leverage and expand on these existing efforts to include more fields and provide better data systematically, as well. Hopefully, more training institutions, funders and future employers will also heed this call and appropriately take on this responsibility.

Postdocs themselves have made some grassroots efforts to answer some of these questions and provide data relevant to the biological sciences. A survey is currently under way in Canada seeking to survey all Canadian postdocs, both at home and abroad.

The UCSF Postdoctoral Scholars Association, which formed in 1995, undertook a survey from March to June of this year of all Ph.D. and M.D. postdoctoral fellows at UCSF. Initial results from both of these efforts are available on the web through *Science* magazine's Next Wave site, and both groups hope to have full reports out this fall.

The vast majority of Canadian postdoc respondents are in the sciences (with a few from the humanities) and there are similar trends reported in both surveys, but I'll limit my remarks to a brief look at the UCSF preliminary results. From their mailed questionnaire they collected responses from over 430 postdocs (a 40% response rate). This revealed that the mean age of the UCSF postdoc is 32 years. Half of the respondents reported that funding was made directly to them and it is further concluded that varied funding sources have produced varied levels of benefits and money for research and travel expenses. On average a postdoc at UCSF works for a principal investigator who supervises 2 to 8 postdocs (mean of 5) and zero to 4 graduate students (mean of 5). Also, about 20% of respondents reported prolonging their postdoctoral position because of difficulty in finding other employment.

Even before improved data is fully available, it is possible to make some recommendations for future training needs and desirable reforms. We suggest:

1. The masters degree should be reinvigorated as an education and training option.
2. That more funding go directly to students and to postdocs so they can seek increased flexibility in their training.
3. Closer interaction must be sought between universities and all types of employers in establishing training agendas and providing training settings. This could include having scientists from non-research careers serve on dissertation advisory committees. This would allow students to broaden their perspectives and networks early in training before critical decisions points in career paths are reached.

4. Industry needs to reevaluate its role and responsibility for training. There is much talk nowadays of employees needing to be life-long learners. This has always been a fact of life for scientists -- but they cannot be expected to be life-long students and postdocs, at some point they must be given the opportunity to be trained on the job.
5. Better evaluation and long-term tracking of all trainees should be institutionalized and standardized at the university, local, regional and national levels.
6. An examination of the current federal funding portfolio needs to be carried out to determine more precisely where investments are being made and if there is a need to shift resources.
7. Postdocs should be given a formal status and identity within the institution where they are training. Clear training goals and outcomes evaluations should be a part of postdoctoral fellowships. And stipend levels and benefits need to be in line with their stage of life needs (e.g. inclusion in retirement savings plans and access to child care).
8. There is currently a large cohort of Ph.D.s in training and in postdoc positions who will not benefit from reforms. These individuals need to be given a high priority as efforts are made to increase job opportunities for young scientists. Barriers to career transitions into other fields such as secondary science education need to be reduced, and more efforts need to be made to keep quality research options available.

If we want to continue to attract the next generation of the best and brightest into science, we need to show them that a diverse array of good careers awaits them along with an inclusive community dedicated to the future of science and to the science futures of their students.

9 The Genome Frontier: Reassessing the Role of the Ph.D. in a Changing Biotechnology Industry

Michael Cronin

I am obviously the other side of the coin here -- in the industrial sector. Happily, like Edward Penhoet (Chapter 5), I have kept a foot in the academic sector. I have an academic appointment at the University of Virginia and I continue to teach there every year. I often get the students into closed sessions where I have heard a lot of their problems, one-on-one.

What I want to do here is a little different. I want to address, from my point of view, a couple of the ideas you have already read. Then I'll turn to an idea for a large science project that California and the U.S. could take the lead on. Finally, I will address briefly just what the manufacturing side of biotechnology is like, representing the view of a more mature company.

The biotechnology industry is 20 years old. Whether it is Chiron or Genentech, hopefully we are beyond the adolescent confusions and are now productive young adults. The start-ups undertaken by a number of visionaries 20 years ago have resulted today in roughly $10 billion in sales, export surpluses, 100,000 employees, and reasonable growth potential. We have over 1,000 smaller biotechnology companies. This creates a demand for the best of our talent, the top fifth percentile from the science and engineering sides in our schools. Moreover, these start-up companies tend to hire young people, which is good news for the

---

Michael Cronin was Senior Scientist at Genentech. This article is based on remarks delivered at the October 2, 1996 symposium, "Meeting the Challenge: Graduate Education in the Biological Sciences in the 21st Century" held at the University of California, San Francisco.

best students. The article by Penhoet (Chapter 5) does an excellent job of describing the range of opportunities in biotech for young scientists and engineers. I could take his figures and go through the same discussion he gave with the conviction I have gained over ten years at Genentech. I would only change one part. When Ed started at his dad's store, I was just beginning to crawl. My first job was at a liquor store in Hollywood, California as a delivery boy, then I was store manager for four years before I was legally allowed to drink liquor. That job helped pay for my first year of undergraduate studies.

I can capture in a word what I look for in a young recruit: *renaissance*. Rather than specializing to the nth degree, my concept of the perfect person to come into my office and interview for a science or engineering job is someone with a broad perspective. The candidate absolutely nails a research area as a graduate student and postdoc, has achieved a high level of impact through publications or other standard criteria, and has received outstanding letters of recommendation. But I also want to see other attributes, and those are what are going to capture my attention.

I want to address anyone's concern that when you put an application into certain companies, they go into a black hole. They do. We get over 500 unsolicited applications a month at Genentech. That is to say nothing about the hundreds we get for formal advertisements in the standard media. It is a true deluge. We don't have hundreds of skilled people to read through the applications and pick out the best and the brightest. It is a real dilemma -- and can be a black hole.

Genentech and other companies in the biotechnology industry want the best and brightest. We want the academician to train those individuals to acquire the most advanced knowledge very quickly and to do something with it, whether as an academic researcher, someone in industry, a teacher, or a scientist working to develop public policy. I obviously have certain criteria. I would love to find the biologist who has succeeded in a biological discipline, who has taken an elective in engineering, who has taken an ethics course, a course in patent law, or a course in economics. Those are the people who, over the long run, are going to be the most adaptable, the most successful in my environment. There is certainly plenty of room for the specialist, who will always be the technician and often the prime mover in a discrete area, so long as that technical or conceptual class survives.

I always ask about a candidate's failures. A lot of young people can't describe a significant failure in their academic life, yet this is a very significant part of how I judge how that person is going to respond to real life situations. How resilient is the person, emotionally, intellectually, and so on?

I would like to address two issues more formally. I will introduce them by mentioning a full-page ad Genentech put in the *Wall Street Journal* this springtime. The ad basically says, "Work Those Genes," and shows the rear pocket on a pair of jeans with the double helix embroidered on it. The ad announces the celebration of Genentech's 20th anniversary and it announces what we do. We discover genes, ideally ones that are commercially and medically important, and "work them" -- work them up, manufacture them, bring them to the nth degree of purity, get them into the salesman's hands, train the physician how to use them properly and, hopefully, sell a product at the end of the day. Our jobs require very high tech skills and provide a great personal experience.

The two issues I wish to discuss are the following: first, there ought to be a long term national research goal to assign function to the orphan genes discovered through the various genome projects. Reinvigoration of physiology graduate training is necessary to accomplish this. California institutions ought to lead this effort. The second issue is the supply and demand of employment opportunities. As the biotechnology sector converts to large scale manufacturing, the industry's growth may be limited by the availability of well-trained production scientists and engineers.

Regarding the first issue, we realize that a first order description of the genome of humans and several other species is imminent.

This gene-rush has yielded not only vast basic knowledge and the potential for products that will aid economic growth, but also a dilemma. These brilliant breakthroughs in molecular biology are obviously constrained by a limited understanding of the genes' functions in the organism. The industrial bet is that this gold-rush to sequence will produce one or two blockbuster genes that can be quickly recognized to reward the enterprise.

Let me provide two analogies of this issue. The acquisition of the genome data is like finding the Rosetta Stone, but without knowing Greek. With the genome, the genetic code and gene separations are defined, like an alphabet and words, but the meaning of the words, the grammatical rules and colloquialisms, are lost to us.

Another analogy: President Thomas Jefferson assayed and calculated the potential of the vast Northwestern America through an inspired investment in the Lewis and Clark survey. However, it took generations to acquire the property and then to create the infrastructure necessary to transform the riches of the West into public benefits for the citizens of the United States. The frontier of the genome is similarly being discovered, and again, we are in position to secure the basic rudimentary knowledge. But are we well positioned to transform it to practical purpose?

The allure of molecular biology and the race to capture the value of gene sequences has seduced many in the commercial, academic, and federal policy sectors. Indeed, graduate and postgraduate training, as well as basic biological research funding over the recent past, has been pegged to genetics and molecular biology, at the expense of disciplines devoted to understanding a gene array operating in the organism in health or sickness. While we should be poised to unleash well-trained systems biologists or physiologists on this genetic windfall, those fields have neither been maintained nor modernized.

I deeply admire and celebrate the advances in genetics and molecular biology, and hold that our universities ought to continue to strengthen these programs. However, the productivity of these discoveries will be ultimately validated in the context of the living organism. Accordingly, the University of California and other leading research universities ought to reinvest in the discipline of physiology in order to gain this next dimension of biological knowledge.

My recommendation is the following: that the assignment of function for the orphan genes ought to be a dominant plank of our national research platform in the 21st century. Such a breathtaking goal is multidisciplinary, playing to the strength of the research university, from biology to information technology to law and public policy. Such a vision derives from but reaches far beyond the more modest human genome project. It should be sufficiently tempting to both academic and commercial sectors that they would seek ways to partner in this great enterprise. It should trigger the renewal of systems biology. Such a vision could be articulated and led by California.

To restate the second issue: as the biotechnology sector begins to balance its discovery research with the need for manufacturing, industry's growth will be limited, in part, by the dearth of production scientists and engineers.

Genentech was formed by drawing heavily from UCSF and Stanford faculty and postdocs to its discovery research base. A year later, when Genentech had to mount a manufacturing campaign in 1979, we hired experienced industrial scientists who were interested in the nascent recombinant DNA technology. We also developed a close relationship with MIT and commercial engineering firms to develop the manufacturing systems technology. Since then, the pattern has been to recruit the best biochemists, cell biologists, and engineers directly from the university, and then train them internally as manufacturing and process science specialists. The large industrial scale, the trade secrets kept within the company, and the art or alchemy of recombinant DNA production have required that the apprenticeship typically be at Genentech.

While the Chirons, Genentechs, Genetics Institutes, and Amgens of American biotechnology now have seasoned manufacturing staffs, what about the 1,000-plus smaller biotechnology companies who need to bring a product idea to a biological test-of-concept? Where will they find the manufacturing capability and expertise? Some will be lucky enough to recruit veterans from the large biotechnology companies or from business alliances with big pharmaceutical companies. Others will have to find a way to survive economically long enough to develop their own capability.

How can graduate education contribute to ensure competitive success? While there is proprietary know how, none of this is black magic. Many companies have learned to produce recombinant products that are accepted by the most demanding regulatory agencies. Universities have set up pilot plants or interdisciplinary programs to meet this challenge. There may be the gift of time in this case. Some of my Genentech colleagues estimate that the shortfall on the scale manufacturing side for most of these start-ups could be 5 to 10 years away.

My recommendation is the following: the university graduate leaders ought to convene a meeting with commercial biotechnology manufacturing mavens to gauge the problem and to lay out a remedy if necessary.

Finally, it is in the best interest of all of us to help university graduate training programs ensure the economic success of commercial biotechnology. This requires constant investment, aggressive and healthy scrutiny, and communication between industry and academia, as well as timely action. I am pleased to see industry's and academia's interest today, and trust in their leadership to secure the quality of graduate education in the biological sciences.

Discussion

Paul Lick, Dean of Biological Sciences: I am a little worried. I remember, in the '70's, being excited by the prospect of numerous biology jobs that were going to be available in connection with the environment, jobs that would help save the environment. Those jobs never appeared. Now, as is evident in our discussion, a new and exciting frontier is appearing. But I hear Dr. Austin saying that her letters to the biotechnology concerns went unheeded. Dr. Cronin explains that 500 letters a month drop into a black hole. What is the connection between these contrasting facts?

Michael Cronin: I was, I think, reflecting the reality. As a responsible individual you would prefer to have the staff personally answer every legitimate request for a job. That is not the real world at Genentech. My message to the person applying is to find a different route. Find someone in the industry that you are targeting to become either a postdoc in or an employee of. Direct your resume to that person and then call that person the next week and ask whether they received your resume. As for a direct connection between the frustration you heard here from Finley Austin and what actually happens, I don't know. But in your question, you start out as saying that the promise in the environmental sciences has never been realized and now there is renewed excitement in biotechnology. Well, I personally feel that this new excitement is justified. I think there is a huge new domain of knowledge that can be acquired, exploited -- in the proper meaning of that word -- and

This discussion is based on remarks delivered at the October 2, 1996 symposium, "Meeting the Challenge: Graduate Education in the Biological Sciences in the 21st Century."

that this new domain should come from the seedbed of California and drive your jobs. I am hopeful, and that is all I can say. I can't predict the future.

10 California's Future Workforce: Those Who Leave the Laboratory

John Barton

About 20% of scientific Ph.D.s leave science -- some to law, business, or journalism schools (see Figure 1). These people are certainly a loss to the scientific professions. Are they also a loss to society?

As a teacher of law, I believe that the answer is clearly no. We have an enormous social need for people who combine solid science with a profession such as law. We may want to consider

Figure 1

Primary Employment of Scientists and Engineers

	Scientists	**Engineers**
Teaching and basic research	44%	22%
Applied R&D	18%	41%
Management R&D and non-R&D	16%	21%
Other	22%	16%

Source: National Science Foundation, *National Patterns of R&D Resources: 1992*, as reported in D. Goldstein, "After the Big Crunch," *Washington Quarterly* (Summer 1995).

John Barton is George E. Osborne Professor of Law at Stanford University. This article is based on remarks delivered at the October 2, 1996 symposium, "Meeting the Challenge: Graduate Education in the Biological Sciences in the 21st Century" held at the University of California, San Francisco.

Figure 2

Total R&D by Performer (California, 1993)

	Performers ($ Mil.)	**% of California**
Federal Government	1,785	5.3%
Industry	26,541	78.7%
Universities/ colleges	2,380	7.1%
FFRDCs	2,499	7.4%
Nonprofits	516	1.5%
Total	33,721	100.0%

Source: National Science Foundation, *National Patterns of R&D Resrources: 1995 Data Update*, quoted in Center for Science and Technology (AAAS), *The Future of Science and Technology in California: Trends and Indicators* (May 1996).

new approaches to educating such people, and serious policy research is an important part of such educational efforts.

Perhaps the best way to make these points is to describe what some of my former students are doing in the "real world." Some have concentrated on intellectual property as patent prosecutors or as litigators. Others have concentrated on venture capital, working with new entrepreneurs to help them understand how to structure a business and sometimes assisting them in finding financing. Still others work with the start-ups that are a little further along, assisting them in structuring strategic alliances or in carrying out an Initial Public Offering. And some work with the regulatory process, assisting the government in dealing with new ethical and policy issues, or devising modifications to the regulatory system.

Figure 3

Sample Policy Issues Deserving Study

Intellectual property system

- What is the appropriate level of coverage and rights conferred?
- What is appropriate allocation of rights for publicly-supported research? For other university research?

Regulatory questions

- Defining appropriate standards for consumer and environmental safety
- Harmonization mechanisms

Biotechnology beyond the biotechnology industry

- Agriculture
- Medicine

There is no question in my mind that these people are providing "value added" and that their contribution to society requires a combination of scientific understanding and of legal understanding. Every one of these persons is different, and it is hard to generalize about appropriate training. But I see three issues that should be considered in designing the training: the choice between the Ph.D. and the masters degree, the role of interdisciplinary courses, and the role of policy studies.

It is not clear that these people require a Ph.D., even though that degree has become an important credential in the policy community. What they need is very solid scientific literacy -- an ability to talk to scientists comprehendingly and to read the relevant literature, an absence of fear of equations, and a comprehension of the process by which the scientific truth evolves. This does not require the specific substantial research apprenticeship that is the heart of the Ph.D., but it does require serious, graduate-level study. Hence, an M.S. degree seems very appropriate for such people.

An interdisciplinary course is a sensible component of such a program. Professors Greely of the Stanford Law School and David Botstein of the Genetics Department and I taught a course in "Law and Biotechnology" in spring 1995 and plan to do it again this coming spring. Our strategy was to explore the various legal issues as we tracked a hypothetical innovation (a new receptor related to HIV) from discovery in a university laboratory to actual marketing (and tort suit), with the obvious intellectual property, strategic alliance, clinical trail, and regulatory approval issues along the way. Our class included biology, law, and medical students -- and we gave mixed teams of students specific problems along the way. It worked very well, and I'd strongly recommend this model as a way to give the members of each discipline insight into the differing perspectives of the other discipline and into the realities of research and law in the economic world.

It is not, however, the way to build the expertise for serious policy analysis. What we need here are serious research projects -- so that the person does an apprenticeship in the actual process of interdisciplinary legal (or business or journalistic) work. These are, I believe, genuinely professional projects and better taught in the professional school to students with a scientific background than taught in a joint degree framework. I have been very pleased with some of the opportunities I have had to supervise such projects: on the patent issues posed by proteins that differ only in biologically unimportant amino acid substituions, on the standards used by the U.S. Department of Agriculture in reviewing field trials of genetically-engineered organisms, on the way in which vaccine insurance legislation for HIV would differ from that in place for

DPT, and on the current pharmaceutical regulatory harmonization efforts. I'd go further: as a society, we need serious policy research in a variety of these legal/scientific areas, and the relatively disinterested milieu of a major university is one of the most important places for such research. Education and research are wisely combined in policy as well as in science.

Section III

Meeting the Challenge

Richard Attiyeh

This section's focus is on the nuts and bolts of the issues we are discussing. The list of contributors is composed of people who are close to the front line in the teaching of doctoral students in biological sciences. Each of us will discuss briefly our views of how we are meeting the challenge from the perspective of our own positions and our own institutions.

Richard Attiyeh is Vice Chancellor for Academic Affairs at the University of California, San Diego.

11 The Perspectives of a Graduate Dean

Richard Attiyeh

Having served as a graduate dean for a number of years, I get a little nervous when people start talking about re-engineering or re-inventing graduate education. What we have in the United States at this time is a system that has been enormously successful and continues to make a major contribution to the nation's future economic well being. Nevertheless, it has been characterized as bloated, hopelessly out of touch with reality, and inhabited by exploitative barons who pursue their self-interest at the expense of their students. I categorically reject this caricature of graduate education. Not only is it incorrect, it encourages over-reaction to legitimate matters of concern. Consequently, my plea is that in trying to improve this system, we remain mindful of the inevitability of unintended consequences and avoid throwing out the baby with the bath water.

This may sound defensive, but I am not in a state of denial. Most of the issues on the table here have been the subject of many a graduate dean's sermons over the years. One only has to look at reports of the Association of Graduate Schools and the Council of Graduate Schools over the past decade to become convinced that we have not been living in blissful ignorance and are more than willing to face up to the hard issues.

There is no question that there are problems with graduate education that require our attention. There always have been and there always will be. That is the price we pay for living in a dynamic world. My hope is that our critics and we ourselves will

---

Richard Attiyeh is Vice Chancellor for Academic Affairs at the University of California, San Diego. *This article is based on remarks delivered at the October 2, 1996 symposium, "Meeting the Challenge: Graduate Education in the Biological Sciences in the 21st Century" held at the University of California, San Francisco.*

proceed cautiously, recognizing that one of the beauties of our decentralized, self-governing system is that it has remarkable flexibility and adaptability.

Most of the issues before us concern the interaction of the changing job market that our graduate students will face with the character of the education they are being provided by our graduate programs. Here are some of the questions we need to answer:

- Will the anticipated decline in the rate of growth of the research enterprise require a decline in the ratio of graduate students to faculty?
- Is the nature of work required of Ph.D.s changing as more of our graduates take jobs in industry?
- Are our graduate programs providing adequate career advising to their students so they can make wise career choices and put their time in graduate school to the best possible use?
- Are we providing our students with the breadth of training needed to survive in an uncertain and changing job market?
- Are we providing our students with the breadth of training needed to survive in an uncertain and changing job market?
- Are we providing our students with the best return on their investment by keeping time to degree as low as possible consistent with their receiving a quality education?

Let's assume that some or all of the answers to these questions require change in our graduate programs. Who then should be responsible for deciding what changes are needed? Will these changes occur "naturally", or will changes in public policy be required?

It is my contention that society will be best served if we leave the reform of graduate education to our individual graduate schools. Government agencies and national organizations simply do not have the knowledge base to decide what is best for graduate education. More importantly, the competition among graduate schools for the best students and the common sense of well informed applicants to graduate school are sufficient to ensure that our system of graduate education will adapt to the changing society it is intended to serve.

Why do I take such an optimistic view of the ability of graduate education to modify itself? The answer is simply that I have observed a widespread interest at both departmental and institutional levels to undertake substantial change in response to their rapidly evolving environment. Let me cite a few examples.

The University of California is adopting a new budget policy that eliminates a long standing incentive to expand graduate enrollment relative to undergraduate enrollment. This will make it financially more attractive for departments with what may now be excessively large graduate programs to devote more of their efforts to undergraduate education. The University of California is also undertaking a revision of its long-term enrollment plan to ensure that its mix of bachelor's, masters, and doctoral programs is responsive to the anticipated societal need.

Among California universities, Stanford, UCLA, and UC San Francisco have all instituted sweeping changes in the organization of their biological science programs in order to increase the breadth and flexibility of the training they provide. Moving in a different direction, the Department of Molecular, Cellular, and Developmental Biology at UC Santa Barbara has reached an agreement with Amgen corporation that will enable the company's scientists to serve as dissertation supervisors in the department's Ph.D. program.

Elsewhere, I am aware that SUNY Buffalo, Case Western Reserve, the University of Maryland, Northwestern University, Rutgers University, and Washington University have all initiated new or reorganized existing biological science graduate programs with a view to avoiding premature or excessive specialization, and in recognition of the fact that most of their graduates will not end up in faculty positions at research universities.

More than a few universities have established innovative masters and/or certificate programs in response to the need for management training by scientists and engineers who assume managerial responsibilities. On my own campus, for example, our school of international relations has started a new international technology management track in its masters program, and our University Extension has a very successful certificate executive program for industry scientists and engineers. At Berkeley, the Haas School of Business has also set up a management of technology certificate program for science and engineering students.

On a national level, the Association of Graduate Schools (AGS) of the Association of American Universities has taken an active interest in these issues. Table 1 summarizes the responses of the 50 graduate deans responding to an AGS survey of its 60 member institutions conducted last year.

Table 1

Association of Graduate Schools' Survey (1995)

	Improving Career Advising?	Increasing Curricular Breadth?	Reducing Time to Degree?
Yes, institution-wide	18	13	23
Yes, at dep't level	21	20	11
No or don't know	11	17	16

Table 2

Association of Graduate Schools' Survey (1996)

Issues Most Frequently Considered at the Institution-wide Level	Issues Most Frequently Considered at the Department Level
Time to Degree New Interdisciplinary Degree Programs Size of Graduate Program Tracking of Ph.D. and Postdoc Placement	Curriculum Size of Graduate Program Tracking of Ph.D. and Postdoc Placement Communication Skills

A more detailed AGS survey conducted this year showed that of the 39 institutions responding, 34 were addressing Committee on Science, Engineering and Public Policy (COSEPUP) report issues institution-wide and of the 235 departments or schools that responded individually, 184 were addressing such issues. Table 2 shows the issues most frequently being considered.

What these examples make clear is that there is a lot more going on than most people realize. And it is happening at the right level. A good deal of the rhetoric about graduate education seems to suggest that we need more directive national policy to guide the education of scientists. Instead, we need to preserve institutional autonomy and initiative. Only through the kinds of local experiments that are currently taking place can we find out what changes constitute real improvements. I can assure you that proven good ideas are quickly copied; and to help ensure that this is true the AGS has set up a worldwide web site reports on successful innovations around the country: http://www-ags.ucsd.edu/ags.html. Although I believe it is appropriate that graduate schools are taking steps to increase breadth, improve career advising, and provide more professional skill training, I sometimes fear that well-meaning reformers will try to turn the Ph.D. into a glorified MBA or a J.D. degree, as was recommended in the September 22, 1996 *New York Times* magazine section.

In the current climate of concern about doctoral education, some people seem to forget that the Ph.D. is first and foremost a research degree. "As such," says the AGS Statement on the COSEPUP Report, "it will continue to serve well both those who pursue research careers in academia, industry, and government and those who become faculty members in primarily teaching institutions, managers in technology-based industry, and policy makers in governmental agencies. The crux of Ph.D. training is the research apprenticeship. It is not, however, the specific, sometimes very narrow, knowledge gained in the research apprenticeship that is most important. The uniqueness of Ph.D. training is learning the process of identifying, defining, and solving a problem. This broad analytic capability is what makes Ph.D.s so attractive to such a wide range of employers."

In conclusion, let me summarize my position. Graduate education in the United States is fundamentally sound. It is a decentralized system that is adaptable and flexible. In response to rather substantial changes in the world around us, a considerable amount of experimentation and change is taking place at the grassroots level. If left to its own devices, graduate education will evolve in scale and character, as it has in the past, so that it will continue to serve well the interests of society.

12 Meeting the Challenge: Graduate Education in the Biological Sciences in the 21st Century

Maresi Nerad

I will discuss a national study, "Ph.D.s -- Ten Years later," which Joseph Cerny, Vice Chancellor of Research and Dean of the Graduate Division at the University of California, Berkeley and I are presently undertaking. This study examines the career paths of some 6,000 Ph.D.s in six disciplines -- biochemistry, computer science, electrical engin-eering, English, mathematics, political science -- from 61 U.S. doctoral granting universities who received their degrees between 1983 and 1985. All nine University of California campuses are participating. About 1,000 biochemists are included in this study. Compared to the total population who were awarded doctorates in these three years, our survey population, including biochemists, comprises 57% of the total number of Ph.D.s awarded.

Project Overview

The major goals of the study are to:

1. Develop an understanding of the outcomes of doctoral education in terms of program satisfaction and subsequent employment. We want to move away from anecdotal evidence towards empirical facts and an accurate employment history for our doctoral graduates. By doing so we can look at issues of

Maresi Nerad is Director of Graduate Research, Graduate Division, University of California, Berkeley. This article is based on remarks delivered at the October 2, 1996 symposium, "Meeting the Challenge: Graduate Education in the Biological Sciences in the 21st Century" held at the University of California, San Francisco.

underemployment and unemployment for the surveyed cohorts in a national context.

2. Link the Survey of Earned Doctorate data (such as financial support, etc.) with doctoral educational experiences and career outcomes.
3. Evaluate the (lengthening) postdoctoral experience by discipline. We want to determine the average length of postdoctoral appointments.
4. Provide career perspectives for graduate advisors and potential graduate students. (We are aware that the job market has changed since the mid-1980s. Major trends, however, have not altered.)
5. Validate the usefulness of the information regarding the employment plans of doctoral students provided in the Survey of Earned Doctorates.

Methods

In addition to determining and analyzing the career paths of some 6,000 doctoral recipients by means of a mail survey, we are conducting in-depth interviews with at least 120 people -- 20 from each discipline. We are combining quantitative and qualitative research methods. We want to learn more about the reasons *why* individuals have chosen a particular career path. Further, we want to learn about specific employment patterns of women and minority doctoral recipients, and analyze career patterns by variables such as program size and nature of doctoral support. For example, do people with multi-year NSF fellowships have significantly different career paths than those without such a prestigious fellowship? And lastly, we want to provide a comprehensive retrospective evaluation of many aspects of doctoral education across these six fields and 61 universities.

Questionnaire

At the end of this month we will have mailed the fairly lengthy questionnaire to each of the 6,000 participants. We developed this questionnaire while undertaking 54 exploratory interviews with individuals from our survey population from the selected disciplines living in various parts of the U.S., and we received valuable input from the study's advisory committee, the University of California, Berkeley Survey Research Center, and from a number of other expert individuals. We refined the questionnaire through pilot tests with 18 non-participating Ph.D. volunteers, both at the University of Washington in Seattle and in the Bay Area.

The questionnaire has six sections, each inquiring about a particular aspect of the post-Ph.D. experience and the doctoral program itself.

1. The first section deals with the individual's career path from the time he or she received the Ph.D. until the present. This includes a number of questions about postdoctoral appointments. It also asks what factors were important in the individual's decision to accept the first (non-postdoc) and current position. In addition, this section asks about job satisfaction at the present time.
2. The second section inquires about how the individual conducted his or her search for the first (non-postdoc) job.
3. The third section is a retrospective evaluation of the individual's doctoral program.
4. The fourth section asks about the individual's opinion regarding the usefulness of the doctoral degree.
5. The fifth section requests demographic information about the individual and his or her spouse or partner. We have in our survey population a generation of dual career couples who may even work on opposite sides of the continent. Therefore we included a number of questions about the partner's employer's location, the type of work, and about children.

6. The sixth section solicits recommendations for our current doctoral programs and present students.

Where are Our Ph.D.s?

After an extensive, laborious search for current addresses, we have found 95% of the addresses for our survey population. We know that not all of these addresses are correct, but we used every conceivable means to validate them. Even so, we are missing current addresses for 309 people. We then divided the U.S. into 5 geographical regions and subsequently found an interesting clustering of people by disciplines. Let's look at the concentrations of biochemistry Ph.D.s (see Figures 1 and 2).

Biochemistry

Our 927 biochemistry Ph.D.s who live in the U.S. are primarily concentrated in the eastern U.S. (39%), a smaller percentage (24%) live in the West, and 21% live in the Midwest. Biochemistry has the second largest proportion of individuals living in the East. In contrast, our 871 electrical engineering Ph.D.s who live in the U.S. are concentrated, not surprisingly, in the western U.S. (40%); a smaller proportion (31%) reside in the East. Of the 348 who live in the West, 84% live in California, and a substantial proportion of these can be found right here in the Bay Area, in the Silicon Valley.

In order to understand the mobility of the biochemists, we had to compare the location of their present residence with the location of their Ph.D. granting institution. Presently, 39% of the biochemists live in the Eastern U.S., but only 28% received their Ph.D.s from eastern universities. Twenty-seven percent received their Ph.D.s from universities located in the Western U.S., but only 24% of all biochemists live presently in the western part of the country. We therefore can speak of a migration of biochemists to the East (see Figures 3 and 4).

Figure 1

Biochemists: Where do they live now?

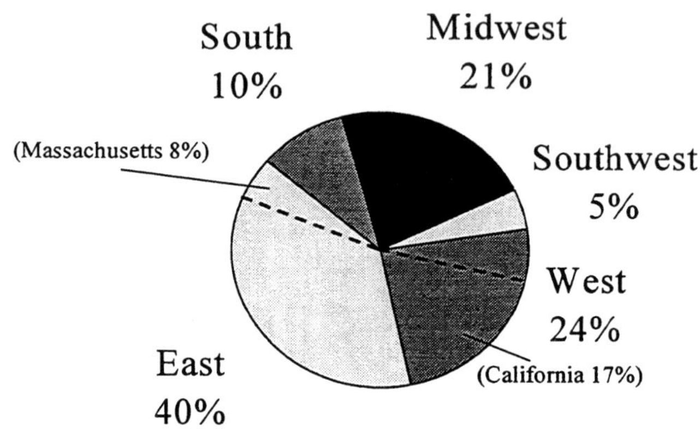

Source: "Ph.D.'s – Ten Years Later" Study,
Graduate Division, UC Berkeley

Figure 2

Electrical Engineers: Where do they live now?

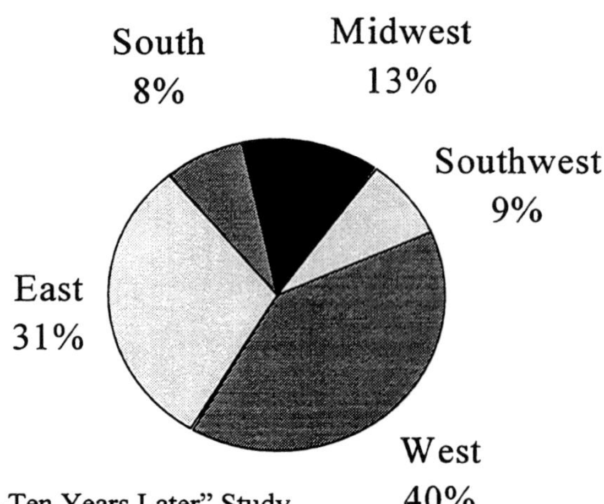

Source: "Ph.D.'s – Ten Years Later" Study,
Graduate Division, UC Berkeley

Figure 3

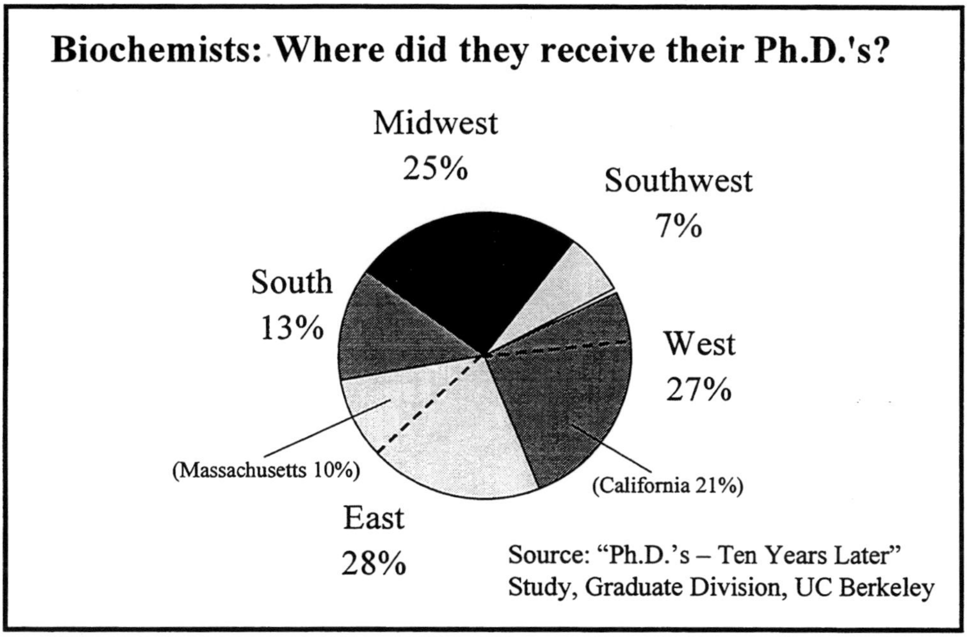

Figure 4

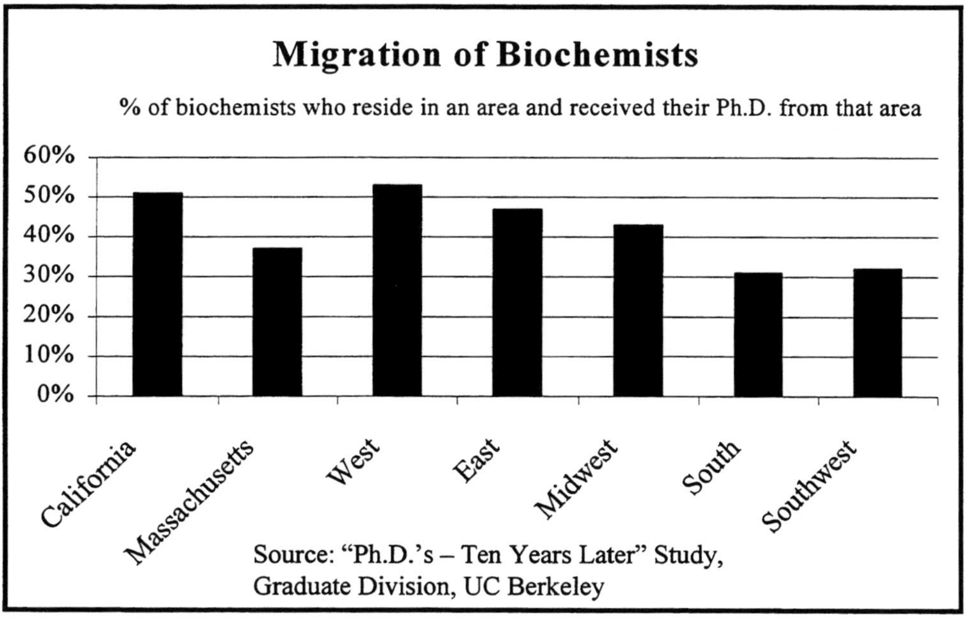

Given the massive bi-coastal concentration, it was not surprising to learn that the choice of work location was not accidental. Interviews confirmed that the geographic location of the work place was a key factor in the decision to accept or reject a job offer. In many cases, people limited themselves to the East or West Coasts in their ultimate employment choices, particularly when selecting their current job. Ph.D.s who grew up on either the East or West Coast or who attended graduate school in either location made a conscious decision either never to leave these areas or to return to them when accepting their current jobs. California Ph.D.s seemed to be particularly prone to reject jobs in any location other than the West Coast.

Historical Comparison

One of the few comprehensive studies about careers of Ph.D.s was undertaken by the National Research Council and published in 1968 by NAS. The analysis of the results of this study looked closely at the geographic mobility of Ph.D.s. We intend to compare our findings with those of 30 years ago where appropriate. We are most eager to compare our findings with those of the 1968 NRC study in terms of career paths. We will analyze whether our Ph.D.s ended up with jobs inside or outside of the higher education sector 10 years after they completed their doctorates. The 1968 NRC study divided the employment history of the surveyed Ph.D.s into four categories: (1) always academic; (2) academic to non-academic; (3) non-academic to academic; and (4) always non-academic. In contrast to the popular image of "the good old days" when all Ph.D.s supposedly got academic jobs (it was, after all, the 1950s and the era of extensive higher education expansion), the NRC study found that 35% of the biological scientists of the 1950 Ph.D. cohort worked in the non-academic sector after 10 years, 24% immediately started to work outside of academia and another 11% joined them later. The present trend towards larger percentages of Ph.D. recipients working outside of the higher

Table 1

1950's Career Paths of Ph.D.s Ten Years Later					
Major Field	**AA**	**A-NA**	**NA-A**	**NA**	**Total**
Physical Sciences	32%	12%	19%	37%	100% (435)
Biological Sciences	48%	11%	17%	24%	100% (558)
Social Sciences	56%	12%	17%	16%	100% (362)
Humanities, Arts, and Prof. School	62%	12%	10%	16%	100% (271)
Total	48%	12%	16%	24%	100%

Source: NAS, Careers of Ph.D.s: Academic Versus Nonacademic. A Second Report of Doctoral Cohorts 1935-1960, p. 14; Washington DC, 1968.
Key: AA=Always Academic; A-NA=Academic to Non-Academic; NA-A=Non-Academic to Academic; NA=Never Academic

degree sector is not a novelty in the history of Ph.D.s and their careers (see Table 1).

What Did Our Interviewed Ph.D.s Think of Their Experiences?

At present we have early findings from 54 individual, one- to two-hour interviews, which were tape recorded and transcribed. Among those interviewed were 14 biochemistry Ph.D.s from the western, eastern and southern parts of the U.S. We posed questions such as the following: What career goals did our Ph.D.s have in mind when they started graduate school? What was their employment history? What did they think of the quality of their Ph.D. programs? What help did they receive in the job search upon completion of their degrees? How useful do they consider their doctoral degree to be for their subsequent career? And what recommendations can they offer our present doctoral programs and students? The answers were revealing.

Career Goals

Most interviewees intended to pursue an academic career at the time they entered graduate school; only a few were not sure what they wanted to do. The biochemists we interviewed most often mentioned their desire to do research as their reason for undertaking graduate study.

Engineering doctorates were most likely to have changed their minds over the course of their doctoral studies and at the end of graduate study frequently decided to pursue industrial careers. At the end of the graduate study biochemists still wanted to do research; half of them wanted faculty positions, the other half did not care in which setting -- academic or non-academic -- they performed research.

Faculty Expectations

We asked our interviewees about their faculty's expectations regarding their students' careers. *All* interviewees, except those from electrical engineering, responded that faculty expected their students to obtain faculty positions after completing the doctorate. The interviewed Ph.D.s resented this attitude on the part of their faculty. Biochemistry Ph.D.s made unsolicited mention of the fact that at the time they were completing the Ph.D., "industry was unacceptable" as a career alternative. "Chemists might go to industry, but not us biochemists," they said.

Career History

In the following, I will present a few examples of the various career paths our interviewees pursued after completing their doctorates. These are not prototypes, but illustrations of the variety of different career paths of biochemists.

1. The first main example is a man who received his Ph.D. from a major public university after twelve years in the program. This time-to-degree has been established by the Survey of Earned Doctorates, and, as we have found, differs in many cases from the actual time to degree recorded at the Ph.D. granting institution. We assume this difference results from the fact that the person completed a masters degree elsewhere and may have taken time off after the masters degree before entering the Ph.D. program. The first individual started his career with a 3-year postdoc at a major research university, moved into a position in a biotech company, became the director of research and development there, and after this company filed bankruptcy, started his own company in the same geographical area because he did not want to relocate.

2. The second person received his Ph.D. from a major public research university after eight years of study. He did a three-year postdoc in his research field at a well-known medical center, then accepted another postdoc position at NIH. His initial appointment was a five-year appointment, and it was extended for an additional three years. At the time of the interview, he had been unemployed one year -- or rather, had worked during that time for no pay at NIH. He is now looking for work in industry.

3. The third person, a woman, started a two-year postdoc after completing the Ph.D. in six years at a major, private research university. She found a job as a senior scientist in industry, and after three years moved to the patent department of the same company and became department manager after two years. After another two years she started her own consulting company

in the field of biotechnological property rights and is now working simultaneously as a research assistant at a major university in order not to lose contact with research.

4. The fourth example is also from a major, private research university and completed his Ph.D. in six years. Immediately after completing the degree he entered a masters program in bioengineering while working as an assistant in the anatomy department. During the course of his doctoral studies he became aware that he loved to make things more than he liked creating new theories, and thus he was very clear that he did not want to stay in academia nor did he want to do a postdoc, but rather he wanted to obtain additional expertise and so undertook a masters degree in biomedical engineering. He landed his first job as a research engineer in a biotech company. After two years he became a senior research engineer in the same company, later became a section head, and is now the director of R&D in microbiology at the same company.

5. Our last example is a woman from a major public research university. After completing her degree, she first had a one-and-a-half year postdoc in pediatrics, then moved to a second postdoc in a medical center, accepted a third three-year postdoc in another university in the department of pathology, and finally obtained a fourth postdoc in clinical chemistry which lasted for two more years. Her series of postdocs was the result of having to move from one part of the country to another in order to accommodate her husband's jobs. Accepting a postdoc was a fairly easy way to ensure employment in her field. Right after she gave birth to her child she continued to work as a postdoc, but since she couldn't keep regular hours she worked without pay for two years. After nine-and-a-half years she found a permanent position in a state public health laboratory and is presently the section chief of the screening and clinical chemistry division.

Overall, we found that the 14 interviewed biochemists had, on the average, two postdocs of a total average length of five years. At present, nine of the 14 work in the non-academic sector. The other five work in the academic sector, and three of these are tenured faculty. I would like to comment, also, that based on the interviews, we found that the tenured faculty seemed to be the least satisfied with their jobs. They loved to do "bench science," but their present positions allowed them to allocate the *least time* to precisely what they loved most to do.

Usefulness of Ph.D.

Let me give you a summary of another area we covered in the interview: the usefulness of the Ph.D. When interviewees were asked, How useful was the Ph.D. for your subsequent career?, they typically responded that they: (1) gained analytical/critical thinking skills; (2) learned endurance; (3) gained mental discipline; and (4) often enjoyed the intellectual stimulus and challenge they experienced during their doctoral studies. When questioned further, they commonly answered that regardless of field, the Ph.D. degree bestowed (5) self-confidence; (6) served as a union card, opened doors, and enabled them to establish credibility with clients or funding agencies. (7) The degree commanded respect. (8) In addition, during their studies many developed contacts which proved to be professionally useful afterwards.

Management Skills and Teamwork

Our interviewees' comments regarding teamwork, management skills, and organizational skills were particularly interesting. We asked them whether they agreed with the present notion that our doctoral programs should teach students organizational/management skills and how to work in teams. We also asked them whether they thought their Ph.D. programs prepared them too narrowly. Surprisingly, we received a vehement NO regarding the narrowness of doctoral programs. Since the Ph.D. prepares people to do research, and since research, per se, requires a certain specialization, this aspect of the doctoral education was considered good. "One has to zero in at some point," they said. But they also thought that university doctoral programs were not the appropriate places to teach business organizational and management skills. Rather, programs should equip students to run a laboratory, to teach and organize people effectively, and to apply for grants. If one expected to be trained in management, they said, then one should either do a Ph.D. in business administration or learn the skills after completing the Ph.D. outside the walls of academia in "real life," not in an artificial university setting. (A number of our interviewees had started their own companies in the mean time!)

Most interesting were the answers regarding team work. Nearly unanimously, those interviewed felt that team work cannot be properly taught at the university since the university promotion and tenure system does not reward people for succeeding in a team endeavor, but for individual research, as in the case of the doctoral thesis. Most extreme is the tenure evaluation where *individual* contributions must be demonstrated. Our interviewees felt that any attempts to teach team work in a doctoral program would basically result in a superficial exercise with no professional consequences in the academic sector, unless there was an accompanying change in the underlying university value system so as to reward collaboration and refrain from glorifying the individual.

In spite of this pointed critique of university values, we received a number of thoughtful recommendations for improving doctoral programs:

1. Cut back enrollment by employing a more selective admissions process, but do *not* eliminate doctoral programs all together;
2. Pay more attention to teaching -- this was the most frequently mentioned recommendation;
3. Establish summer internships with local organizations, private companies, libraries, or publishing houses;

4. Set up colloquia to bring to the campus Ph.D.s -- possibly alumni -- who work in a variety of fields to help make students aware of non-academic career alternatives;
5. Programs should look outward and be more involved in social concerns so that students can learn about the concrete, societally relevant applications of knowledge in their field.

In summary, our surveyed students were satisfied with the education and training they received to become critical/analytical thinkers and competent researchers. They were critical of the lack of training in how to teach, and definitely wished they had been provided with more information on employment possibilities outside the higher education sector.

During the last two years, our universities, and particularly the graduate schools, have made remarkable efforts to implement teacher training programs for the training of future faculty, and we are slowly moving to provide information on employment alternatives outside the university. Hopefully, our faculty will adjust their attitudes and begin to convey to their students that an academic career is not the only successful career path for a Ph.D.

13 Fostering Breadth in Graduate Studies

Craig Heller

I will discuss a problem inherent in graduate education preparing future Ph.D.s for a variety of career options other than academia. That problem arises from increasing specialization in graduate research and increasing narrowness of training that characterizes Ph.D. programs. This has been the result of a steady process over the last decade or two.

It is unavoidable to use one's own experience as a reference point. Like many in my cohort when I entered graduate education 30 years ago, I hadn't a ghost of an idea what I wanted to do. I was able to flounder around a bit in a very fertile environment and look at different options. It is more difficult to have that experience these days. One of the reasons is the growth of departments devoted to increasingly narrow specializations or disciplines. In the article by Kenneth Shine (Chapter 3) it is pointed out that there are many departments in the basic sciences that are only now being recognized as important centers of graduate training. That is true, but many of these departments are really only recent creations.

If you go back 15 or 20 years, you would be hard-pressed to find a department of neurobiology. Neurology, yes, but not a basic science department of neurobiology. You wouldn't find many departments of genetics. You wouldn't find departments of developmental biology. These disciplines were represented within broad-based departments of biological sciences, which still exist. Now we have lots of more narrowly defined, discipline-based departments that engage in graduate training.

Craig Heller is Associate Dean for Research at Stanford University. This article is based on remarks delivered at the October 2, 1996 symposium, "Meeting the Challenge: Graduate Education in the Biological Sciences in the 21st Century" held at the University of California, San Francisco.

The undergraduate junior who is thinking about going on to a career in the biological sciences and is therefore seeking a Ph.D. degree has to make a decision. To what department should he or she apply? This decision is another narrowing point in the training process. It occurs before the individual has completed the undergraduate degree -- even before the student has had an opportunity in undergraduate years to be exposed to different research and research possibilities.

Even in broad-based departments, there has been a narrowing of graduate training reflected in decreasing dependence on course work and decreasing emphasis on breadth requirements in curricula. We all think our graduate students should get right to the bench, get right to work. We tell them to decrease the time towards completion of the thesis. Do a significant piece of work. Learn the latest techniques. Why worry about other areas that you're not going to pursue later on?

I am sure many people can remember that when they were in their Ph.D. program there was such a thing as qualifying exams, and frequently these included comprehensive exams in disciplines other than your own. If you were a biochemist, you might also have to study for qualifying exams in ecology and evolution or in cell development or in some other discipline. When you completed your qualifying exams, you had an amazing breadth of knowledge in your field.

We are the ones who have made the changes in graduate education that have narrowed and focused it, and we have had good reasons for doing so. A major factor has been the drive to make our students more competitive in the job market. It is important in the job market to be one of the top two or three people in a field. The way to do that is to narrow the field in which you are claiming to be an expert. The increasing complexity of technology in our various areas of research has also contributed to this narrowing effect.

For example, in my field, a graduate student can learn nothing more than patch clamping or calcium imaging, and spend four to five years doing a Ph.D. using that one technique. When you have a very expensive, very difficult, and very complex technology to master, you decrease the range and kinds of questions and problems that you are able to explore.

The results of this narrowing process have been fine in many ways for academia, in the sense that we have been able to clone ourselves. At least, it appears that way in terms of research and research productivity. But many people get a little bit frustrated when the new assistant professors that have been hired are unwilling or find it difficult to teach introductory courses. There are issues about teaching functions in academia that are not well served by the specialization that has occurred in graduate training. I have recently heard a very interesting proposal. The suggestion is that we turn the teaching responsibilities in our departments upside down. The senior faculty should all be teaching the introductory courses. They are the ones with the greatest breadth of knowledge but perhaps are out of touch with some of the newest areas. Junior faculty should be teaching all of the advanced graduate level courses which would, of course, focus on their individual areas of expertise. This proposal might make good sense but, still, you might think that someone who has been trained in a graduate program in a good university should be able to teach an introductory course in his or her broad discipline.

I will discuss a possible solution. This solution has already been tried at UC San Francisco in the Program in Biological Sciences described by Keith Yamamoto (Chapter 15). It is one that we have been working on at Stanford over the past year. Graduate training in biomedical sciences at Stanford includes about 200 faculty mentors and about 350 Ph.D. candidates. Until this past year, when students applied for graduate studies at Stanford, they had to submit applications to one and only one of the possible departments where they could get graduate training in the biological/biomedical sciences. This created a dilemma for the student who didn't know exactly what he or she wanted to do. As I said before, this was a critical decision that narrowed down the range of possibilities the student would have for professional development. Even in a broadly-based department like biological sciences, the realm of possibilities is narrow. If a senior undergraduate was applying to Stanford, and that senior was interested in genetics, he or she would read through the options in brochures and catalogues.

For example, perhaps the student wanted to go into corn genetics, drosophila genetics, neurospora genetics, or population genetics. But if the same individual applied to the department of genetics, there would be different opportunities -- perhaps human genetics, mouse genetics, or *Arabidopsis* genetics. There is a problem hidden here. The specialty departments are basic science departments in our medical school. Because they often have graduate student support largely from grant funding, the incoming student must make a commitment early in his or her graduate career to work on a specific research project to obtain a research assistantship.

What we have done this past year is to create one admissions program for all of the biomedical sciences in a combined admissions program. The student that is inquiring about graduate training programs in biomedical sciences at Stanford gets one brochure. That brochure is an invitation to submit one application to the University that is applicable to all of these programs. After having a couple of nice pictures of what it is like to live in the Bay Area with Yosemite and San Francisco nearby, the next thing in the brochure is a long listing of the faculty by areas of interest. One faculty member can appear in the listing more than once. The student immediately gets the view that faculty interested in a given area of research may come from a variety of departments, which reduces emphasis on individual departments or programs. The brochure describes different departments that are called Home Programs. Faculty in those departments are described in a little more detail, including some characteristics of their research programs.

There is a very interesting table in the back of the brochure that emphasizes that the student can find an intellectual match with a faculty member and lab group that share a variety of interests. It displays each faculty member's various interest areas, from biochemistry to population biology. If a student is interested in neurobiology and comparative aspects or evolutionary aspects, the table shows any faculty member who has checked off relevant columns such as evolutionary biology, neurobiology, and maybe physiology. The brochure is our effort to represent the breadth of academic disciplines and possibilities that exist at Stanford.

How does the Stanford admissions program work? In the application, the student indicates first, second, and third preferences for Home Programs. All of the applications are shared between all of the Home Programs, each of which has its own admissions committee. These committees rank the candidates and communicate with each other as to whom is going to offer admissions to which student.

Offers of Admissions or a Home Program are sent out centrally. Students are informed that admission includes access to *all* of the biomedical programs available at Stanford. When the student arrives on campus, the admitting Home Program is responsible for funding the student that first year and the student must undertake a first rotation in that Home Program. The subsequent two or three rotations in that first year can be done in any of the biomedical programs. Neither students nor faculty can make long range thesis commitments until April 1 of that first year.

What are the results? We only have a sample size of one and only one year of experience with this combined admissions program. Application file review went well. It could have been a nightmare, but it wasn't. Selection of candidates went smoothly because there was a high degree of agreement and departments participated cooperatively and enthusiastically in candidate visitation week. Perhaps this resulted from some element of competition between the departments. There was a real incentive to be represented in visitation week and show students what they had to offer.

The yield was very high. Compared to 1994, when we achieved 36% yield on offers to 17% of the applicants, this past year we got 47% yield on offers to 13% of applicants. We think students really like to see such a range of possibilities and value the opportunity to keep their options open. One of the things that we hadn't anticipated, but is now very clear, is the high degree of camaraderie within this cohort of students. We expect this to be a benefit that will carry through their years of graduate training. There is likely to be a lot more communication between these students as they work in the different departments than had occurred in the past. The students are actively seeking rotations outside of their own programs -- which didn't happen before.

Although everything seems to be working well, there is a lurking problem. It remains to be seen how funding arrangements are going to work out. We have all agreed that Home Programs will fund the first year, regardless of what rotation the students pursue. Then, when the students commit to a specific lab for the rest of their graduate work, that lab, department, and mentor must supply continuing funding. We are much less able to predict the funding problems that may lie down the road.

What are we doing about the problem? We are setting up a fund that will provide flexibility in graduate student support. Stanford's President has announced a fund raising drive to create a $200 million endowment for graduate fellowships in the sciences and engineering. This will reduce funding worries regardless of how students vote with their feet.

The problem, of course, is how to fill up a $200 million graduate fellowship fund. We have a strategy that also helps us address another problem. We have problems dealing with intellectual property, especially with small start-up companies that don't want to pay cash for license fees. The university has been loathe to take equity because of concern about creating institutional conflicts of interests.

For example, schools and departments might make decisions on promotions based on the need to enhance the value of the equity that they are holding rather than on the academic merits of the case. To avoid the potential for such conflict of interest, Stanford adopted a policy that, when the university takes equity in exchange for licensing intellectual property, that equity will go into a blind trust. Whenever the fund managers liquidate any of these assets, the proceeds will go directly to the Stanford graduate fellowship fund. No department and no school will stand to benefit directly by the success of one or another company in which the university holds equity.

In summary, we hope that our new combined admissions program in the biomedical sciences will allow students to gain more experience and more breadth before making that critical decision of a thesis project. We believe that this new, more flexible graduate admission system will increase communication between students and between faculty across departments and schools. Essential to its success, however, is the need to develop the capacity to be flexible in the funding of graduate studies.

14 Re-Evaluating Attitudes Towards Non-Academic Careers

Elizabeth Neufeld

In preparation for this paper, I asked the faculty of Biological Chemistry (UCLA) whether they believed that there was a glut of Ph.D.s and if so, whether it called for a change in our educational program.

The discussion was intense, views were expressed in many ways, but there emerged a general consensus. What I heard was that there probably is no glut because our graduates and postdocs eventually get satisfactory jobs in their field of interest, though it may take them longer than they would like to obtain such jobs. There is some attrition -- some individuals may drop out of science (or at least out of research) during or after their training period because they discover that it is not for them, but the level of attrition has not changed much for at least fifteen years.

I heard the faculty say that we should not be trying to clone ourselves, that not all students should be going to academic positions at research universities, and that a number of their students had gone to other positions (industry, patent law, and teaching in non-research institutions). This tolerance for different careers probably represents a change from the views of a even a few years ago. The faculty also stated that we should not change our educational program, which right now is broad and directed towards problem-solving. However, they say we should ensure that our graduates acquire effective communication skills, both in

---

Elizabeth Neufeld is Chairman and professor of the Department of Biological Chemistry at the University of California, Los Angeles. *This article is based on remarks delivered at the October 2, 1996 symposium, "Meeting the Challenge: Graduate Education in the Biological Sciences in the 21st Century" held at the University of California, San Francisco.*

speaking and writing, because those are necessary for whatever field they will enter.

I then asked the director of a large interdepartmental training grant whether training grants are biasing students or faculty towards training for academic careers. He stated that such was not the case, that the NIH wants the training grant to "move the biomedical enterprise forward" and the concept of "biomedical enterprise" is being interpreted broadly to include not only academic research institutions or industry, but also government, journalism, patent law, and science education at all levels including community colleges and even high schools. He also stated that our trainees have been doing very well with respect to obtaining jobs. Like the faculty of Biological Chemistry, he was generally optimistic about future employment opportunities in biomedical sciences, given the rapid advances fueled by the genome project and by biotechnology in general.

I asked our graduate students and postdoctoral fellows to meet and consider the issue. A few weeks later, I received their answer by e-mail. They said they were pleased with the broad education that they were receiving at this time and did not want to make major changes in our educational program, except to be allowed to take additional courses to further broaden their interest in and exposure to other areas.

But their main point was that *they were afraid to speak to their mentors about alternative careers*, lest they be thought insufficiently dedicated to research and stereotyped as second-rate scientists. I confess that I was quite taken aback, because I thought that we had good rapport with our students and that we conversed freely with them on all topics. I checked this with a recent graduate whom I knew well, as he had served as our student representative for several years. He confirmed that students were afraid to speak of alternative careers to their mentors, or even within earshot of their mentors, for fear of being pegged as inferior students. Then I thought of my own experience with a postdoctoral fellow who told me that he wanted to go into academe, and of my concern that he might feel insulted if I advised him of more suitable options.

Clearly there is reticence on the part of both trainees and mentors to be open about the issue of careers. Our students and postdocs tell us what they think we want to hear, and thus prevent us from helping them prepare themselves for careers for which they might be best suited.

This is a problem for which there is a simple solution. We must initiate an honest dialogue that will involve students, postdocs and faculty. It can take place within departments, or be inter-departmental; it can be in the form of small local meetings, or of symposia that would bring in prestigious outside speakers on the subject of career pathways. The message that I would like to convey is that we must see to it that channels of communication remain open on the subject of greatest importance to our students and postdocs -- namely, what is best for their future.

15 Placing the Ph.D. at the Hub of Career Development in the Biomedical Sciences

Keith Yamamoto

I am going to reframe slightly the charge that has brought together the contributors to this conference and volume. First, by necessity, I shall limit my comments on education to the biomedical sciences, as that is the only arena in which I have some relevant experience. Second, I shall consider not only graduate education, but postdoctoral education, too. As you will see, I view postdoctoral research and education as one branch of a multibranched network of education and training that should radiate outward from the Ph.D.

Given this frame of reference, I shall suggest that the Ph.D. should be near the *temporal* center of this radiating network, and that one of my specific concerns about graduate education is that the Ph.D. is now off-center and moving progressively further off-center. I shall propose that the postdoctoral research period should take on new educational responsibilities, some of which are now considered to be in the realm of graduate education and other parts of which are traditionally in the domain of assistant professorship.

My comments on graduate education, then, will be framed against the backdrop of a growing concern -- the time-to-completion of the Ph.D. -- a concern noted in other articles in this volume and an issue with profound impact on everything that we are discussing today. Obviously, extended graduate student training periods have substantial economic implications. Perhaps

---

Keith Yamamoto is Chairman and of the Department of Cellular and Molecular Pharmacology at the University of California, San Francisco. This article is based on remarks delivered at the October 2, 1996 symposium, "Meeting the Challenge: Graduate Education in the Biological Sciences in the 21st Century" held at the University of California, San Francisco.

more importantly, many students perceive a point of diminishing returns, beyond which their career options begin to narrow and become less attractive. The morale of students who train for very long periods commonly ranges from fragile to shattered, and they grow increasingly cynical about whether we are preparing them for their future, or exploiting them for our present.

To confront these concerns, I suggest that we need to reassess the missions, principles, and mechanisms of both the graduate and postdoctoral phases of biomedical education.

Let us begin with the assertion that the mission of doctoral education is to train experimentalists and scholars in the biomedical sciences. I would venture that many might agree with this statement. But now consider what is *not* represented in that statement. *Not* mentioned is a finely honed specialization by the student, or hot and flashy results emerging from the research. There is nothing in the statement about training students to be principal investigators in academic biomedical research. The statement is similarly silent about training students to become biotech company presidents or science writers. That is, the rather bare bones statement that we began with may describe rather well our primary expectations for graduate education *per se*, even though it avoids entirely any consideration of the nature of the research findings, or of the eventual career paths of the graduates.

Now, what might happen if we used our simple mission statement to guide and *limit* our expectations for graduate education? I suggest that we might be able to decrease the time-to-completion of the Ph.D. degree, now ~6.5 years in biology and climbing, by perhaps two years. If we set 4.5 years as a benchmark for time-to-completion, how might we achieve it? In a *New York Times Magazine* (Sept. 22, 1996) article, Louis Menand proposes one way to do it: reduce the Ph.D. to a trade degree, a certification like a law degree. Three years of courses and be done with it. I think that is the worst possible approach. Whatever is done with the time-to-completion, it must be done without sacrificing the standards that define the degree -- scholarship, originality, perseverance. I believe that we can maintain those standards yet reduce the time-to-completion by revising our expectations of the magnitude of the research achievements that represent the Ph.D.

Nobody understands fully why the time-to-degree is increasing as steeply as it is, but it is certainly true that recent technological advances have fueled a remarkable period of discovery in biology and made it feasible for us to challenge our students with questions of unprecedented complexity. Excitedly, the students accept those challenges -- but the demands are great, often requiring technological progress before conceptual issues can even be considered. Thus, our students plunge into complex cloning regimens, development of new strains or animals, design of novel imaging systems and the like. All the while they, and we, hold in front of them the original challenge of that enticing biological question -- and they, and we, expect answers to that question, complex as it is, as a part of the Ph.D. degree. The result is that the time-to-completion increases with the complexity of the questions, and the students stay students longer and longer.

What if we were to rethink the magnitude of the problems that we lay out for our students, or our expectations of the progress they must make toward answering those questions before a degree is granted? What would best define our standards in that context? It seems to me that we should strive to impart three things in our graduate training: biological literacy, research skills, and scholarship. If our students leave our labs with these three tools, they should be equipped to take on the whole gamut of existing problems, to move into fields only tangentially related to their research lives, and to recognize emerging issues of interest and import. This should be our goal. Basically, we want graduate education to foster in our students the same sort of instincts and attitudes enunciated by hockey great Wayne Gretzky: "I don't skate to where the puck is, I skate to where it is gonna be."

How do we do this? To answer that, let's look at these three tools separately, and consider how best to weave them into the fabric of our graduate programs. First, *biological literacy*, in which

I am referring to a working knowledge of fundamental biological principles, coupled with a baseline familiarity with a broad range of experimental approaches, experimental systems, and so forth (see Figure 1). The principles, in my view, are most logically and efficiently set forth in didactic courses, whereas the necessary breadth of familiarity can be designed into the overall program.

By this view, the goal of our courses would be to present the basic principles that are the foundation of all of our experimental approaches -- the principles of genetics, biophysics and molecular structure, and chemistry serve as the main pillars on which our approaches, such as molecular biology, cell biology or physiology, are based. As shown in Figure 1, it is important that these principles and the approaches be interconnected. For example, if a molecular biology program fails to impart an appreciation of physiology (and as Michael Cronin discusses in this volume, many *do* fail), the gene cloners will not know what to do with their hard-won pieces of the genome.

Coupled with expertise in basic principles, graduate students must be presented with a broad spectrum of experimental approaches and systems that provide sufficient familiarity and understanding of numerous approaches so that they can recognize opportunities, choose appropriate directions, and take new approaches without fear. Perhaps the best strategy for presenting a broad spectrum of opportunities for students is formation of interdisciplinary training consortia such as that formed at UC San Francisco in 1988, the Program in Biological Sciences (PIBS; see Figure 1). PIBS encompasses degree programs in biochemistry and molecular biology, genetics, cell biology, developmental biology, neuroscience, immunology, and most recently biophysics; a biomedical chemistry degree program is in the planning stage.

Students apply not to degree programs or departments at UCSF, they apply to the PIBS consortium. Entering students typically

Figure 1

Doctoral Education

Biological Literacy

Didactic Courses:

Principles	*Approaches*
genetics	molecular biology
biophysics	cell-biology
chemistry	physiology

Degree program consortia
(e.g. UC San Francisco Program in Biological Sciences):

- Biochemistry and Molecular Biology
- Genetics
- Cell Biology
- Developmental Biology
- Neuroscience
- Immunology and Microbiology
- Biophysics
- Biomedical Chemistry

Research Skills

- Identify high-impact research questions/issues
- Develop hypothesis and design feasible experiments
- Recognize meaningful findings

Scholarship

- Revise expectations for dissertations (i.e., exhaustive review, publications, overall length)
- Require critique of field, interpretations of findings in context of previous knowledge, perspective

think they know what they want to do, but they commonly discover through laboratory rotations, courses, discussions with peers, or through opportunities emerging from their own research that they are more excited about another field or another approach. A student may enter as a geneticist and leave as a biophysicist, or a student may initiate a thesis project in a given laboratory and carry out major studies in other labs in order to acquire and use the expertise needed to move the research forward. Essentially, the students are presented a broad menu, and told, "here are some core courses with key principles; here is a large faculty with diverse interests and expertise for you to talk to; move from lab to lab during rotations; be assured that you can continue to do so during your thesis work; this is the way science is going to be, so this is a great time for you to avail yourselves of the opportunities to be educated broadly."

So, students who are biologically literate are armed for current research, and prepared for future developments. This literacy must be coupled to the acquisition of research skills -- the second essential component of graduate education (see Figure 1). By research skills I am not referring here to knowing how to use an ultracentrifuge, but rather to three general thought processes required in scientific research: (1) the ability to identify important research questions or issues; (2) the ability to reduce an important issue to testable questions and to design appropriate experiments; (3) the ability to recognize a meaningful finding -- one that suggests future experiments, demands new interpretations of previous data or challenges existing paradigms. These are skills acquired in the research lab through the direct experience of the student, through conscientious mentoring by the advisor, and facilitated by ongoing interactions with peers and other faculty, together with a working biological literacy as discussed earlier. Notably, acquiring these skills does not require spectacular research advances, founding a new field, invited minireviews or cover stories in trendy journals.

The third essential component of graduate education is scholarship (see Figure 1). In this context, consider the doctoral thesis. We know of programs, especially outside of the US, that require a certain number of publications as a prerequisite for the Ph.D. We know of mentors, and students as well, who believe that a real thesis must be a grand tome that is unwieldy to lug around. There are those who measure scholarship as the ability to access every relevant reference from Medline, and to work them all into an encyclopedic introductory chapter. It takes a long time to write a thesis this way. Writing a thesis this way extends the time-to-completion in a manner that is unrelated to true scholarship. Rather, we should demand a thoughtful and critical overview of the field and its overarching issues, an interpretation of the new research findings against the backdrop of what was known before, and a perspective on where the findings may lead. Such a critique, which avoids and rejects the notion of an exhaustive review, embodies the essence of scholarship, and a well framed dissertation represents an opportunity for the student to develop and demonstrate the scholarship component of graduate education.

Students who master these three components (biological literacy, general research skills, scholarship) would, in my opinion, be ready to "skate to where the puck's gonna be". I am convinced that these components could be acquired much faster than the current 6.5 years, mainly by developing graduate programs that maintain a clear focus on the essential components, and by making more realistic our expectations of the magnitude of the research findings.

Clearly, one segment of those emerging from this pool of Ph.D.s would be our future professors and commercial biotech leaders. Will the version of graduate education that I have enunciated generate more Bruce Alberts and more Ed Penhoets? I would say yes, but it would be hard to recognize them without good programs for postdoctoral education, because it is during postdoctoral work that we should be training independent investigators in the biomedical sciences. This simply stated mission for postdoctoral education is notably different from that for graduate education.

Graduate research typically derives rather directly from ongoing studies in the mentor's lab, and save for exceptional individuals, it can be difficult to assess with confidence creativity and independence. I suggest that it is during the postdoctoral period that we should nurture these qualities and expect them to emerge. However, universities and mentors traditionally have not viewed the postdoctoral research period as one with an explicit educational mission. It is notcommonly considered even an apprenticeship, in which it is the responsibility of the mentor or the institution to introduce the postdoc to the various duties of assistant professorship. Rather, the postdoctoral training period seems most typically to be a simple extension of graduate school, albeit commonly with a different mentor and in a different institution, and lacking any of the programmatic coherence of graduate education.

At present, postdoctoral training in biology averages just over three years. I propose that this period should be extended to four years, and that the mentor should assume a direct and active educational responsibility for the trainee. The position would take on elements of an apprenticeship, in which the goals would include demonstration by the trainee of independent research achievement in a specialized area, as well as familiarization with other roles and responsibilities of independent investigators, such as manuscript reviewing, grant writing and reviewing, or supervision and mentoring of students. Implicit in this rather specific mission for the postdoctoral is that Ph.D.s that choose not to become independent researchers would not enter this pathway, and instead would choose other directions in which to refine their training, such as schools of law, journalism, business, education, medicine and so forth. Many of these career paths would also require on the order of four years to complete. Thus, the Ph.D. degree would be close to the temporal center of a multibranched pathway of career choice and development in biology.

What would the "new" postdoctoral traineeship look like? As always, the trainee would choose a specific research topic and an appropriate mentor, and focus on making the strongest contribution possible. But with a widely adopted understanding that this is a four year training period, additional responsibilities and expectations could be added. The mentor would work with the trainee on preparing thoughtful and critical reviews of manuscripts, and in working with students. Institutions would develop postdoc programs and retreats that focus on postdoc research program development and accomplishments, job seeking, grant writing, teaching, the mechanics of running a lab, and other elements of assistant professorship.

Finally, the trainee would be expected to devise and begin to develop an individual research problem, relatively independent of other projects that are aligned with the mentor's studies. Help in this endeavor would be expected from the mentor and others at the institution. Hence, the new postdoctoral traineeship would take on the element of independence that is implicit, but widely unaddressed, in many graduate programs, as well as other functions commonly reserved for the early, difficult days of junior faculty membership.

Certainly, the best individuals, labs and institutions have already been doing parts or all of what is in this proposed scheme for postdoctoral education. Indeed, programs like the Whitehead Fellows and the UCSF Fellows underwrite partial salary, supplies and personnel costs for selected post-Ph.D. scientists to develop their own research programs in an environment protected from the other rigors of faculty membership. But by demarcating specific goals and responsibilities for all postdoctoral training, the nature of the endeavor would change, and would become clearly distinguished from graduate training.

The expectation that the postdoc devises an independent project would do three important things. First, it would strongly boost the confidence of the postdocs to know that they have their own ideas, have begun to develop tools to pursue them, and even have some preliminary results to put into an NIH grant application. Second, it would help to counter the concern voiced by Bruce Alberts that we are making clones of ourselves: creative postdocs will venture far afield of the mainline interests of the mentor laboratory to develop novel ideas. Third, it would better inform the decisions of institutions and funding agencies that are hiring and funding new investigators.

In summary, then, I have suggested that we consider an altered vantage point on two phases of training in the biomedical sciences. For graduate education, we could reduce substantially the time-to-completion of the Ph.D. by rethinking the kinds of discoveries that we expect in a dissertation, by de-emphasizing subspecialization, by modulating the stated expectation of research independence, and by not demanding major research breakthroughs as a defining feature of the degree. By tightening our focus on the development of experimentalists and scholars, we might more efficiently produce Ph.D.s who are better prepared and more appropriately positioned to make decisions about their final phase of training. For postdoctoral study, by slightly extending the training period, we might gain a new measure of independence that allows a faster start for bright researchers, and better provides them with the tools and experience to get started.

The economic impact of these ideas should be positive or, at worst, a wash. Graduate education would be of shorter duration by 35-40%, thus reducing commitments from grants and fellowships for stipends, tuition and fees. Postdoctoral training would be longer by a third, but more students would opt for different training opportunities following graduate school, so there would be fewer trainees. Private and government funding agencies and foundations could recast a smaller number of fellowships as four year awards.

More significant than the economic implications is the reorientation of our view of the Ph.D. degree, and the role of graduate education in biological and biomedical research (see Figure 2). Not long ago, the Ph.D. and postdoc were viewed as the two steps along a linear path between the baccalaureate and assistant professorship. I suggest that the "new" Ph.D. may better serve as the hub from which radiates multiple spokes representing

Figure 2

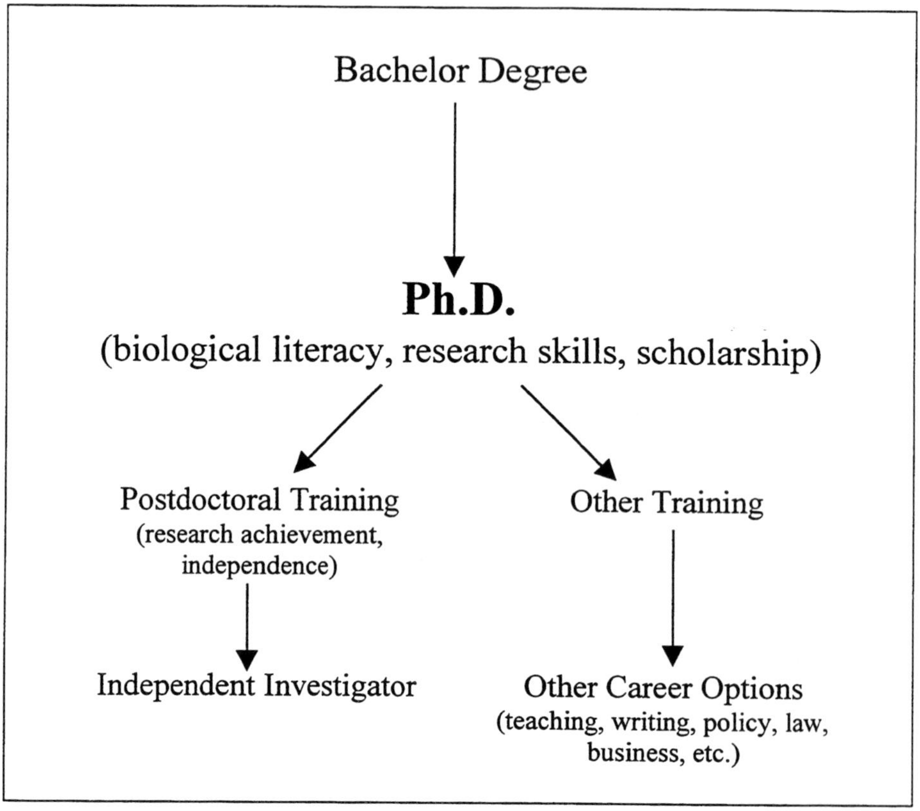

different career tracks, most or all of which require a final, specialized training period. From the Ph.D. would come biological literacy, research skills, and scholarship -- tools that will aid the next stage of decision-making. Not having spent an excessive period in the graduate training phase, students may acquire appropriate training without feeling embittered or exploited. If they choose to go on to law school or business school, they won't be squeezed first for prize-winning or journal cover-garnering scientific contributions. If, instead, they decide to go on to postdoctoral study with the goal of becoming independent investigators, they can expect explicit training toward this goal, and the opportunity and expectation that they will develop a new line of experimentation during that period.

These ideas are simple. I have centered them around the issue of time-to-completion because it is a useful organizing principle and a quantifiable endpoint. The main messages I leave you with are: (1) for graduate education in the 21st century we must present students with broad programs that provide literacy and skills that can be applied across the whole dynamic field in which we now operate; and (2) by adjusting missions and goals of graduate and postdoctoral education, we could place the Ph.D. at the center of a broad range of career opportunities, allowing students to be appropriately trained, and to make wise and timely decisions about their futures.

16 Back to the Future

A. Stephen Dahms

An important aspect of the challenge facing graduate education in the biological sciences in the 21st century can be appreciated by close examination of some of the data presented in The Committee on Science, Engineering and Public Policy (COSEPUP) report (reprinted in Figures 1 and 2), showing the employment of life scientists and chemists 5-8 years after the Ph.D. covering the years 1977-1991. I propose that the future placement of graduate molecular life scientists in the next 25 years will mirror the trends experienced by the Ph.D. chemists over the past 15 years.

In the case of the life scientist data in Figure 1, in the period of 1977 to 1991, there clearly is a decreasing entry into the academic arena (58% of Ph.D.s decreasing to 49%) with a corresponding increase in the numbers entering business/industry (13% increasing to 25%). Chemists over the same time period decreased entry into academia from 32% to 21% and increased entry into business/industry from 46% to 60%. It is suggested that the 1977-91 chemist data may be a harbinger of the future 2006-2020 life scientist data, i.e., the fraction of graduate-trained life scientists entering business/industry in the year 2020 may come close to tripling compared to 1991. It is also suggested that during the intervening 1991-2005 period the numbers of life scientists entering the corporate arena may come close to doubling compared to the 1991 25% level. The needs for, and the migration of these university-trained products into industry should have, or at least can have, significant impact on the nature of graduate training

---

A. Stephen Dahms is professor of Chemistry at San Diego State University and Director of California State University's Program for Education and Research in Biotechnology. This article is based on remarks delivered at the October 2, 1996 symposium, "Meeting the Challenge: Graduate Education in the Biological Sciences in the 21st Century" held at the University of California, San Francisco.

igure 1

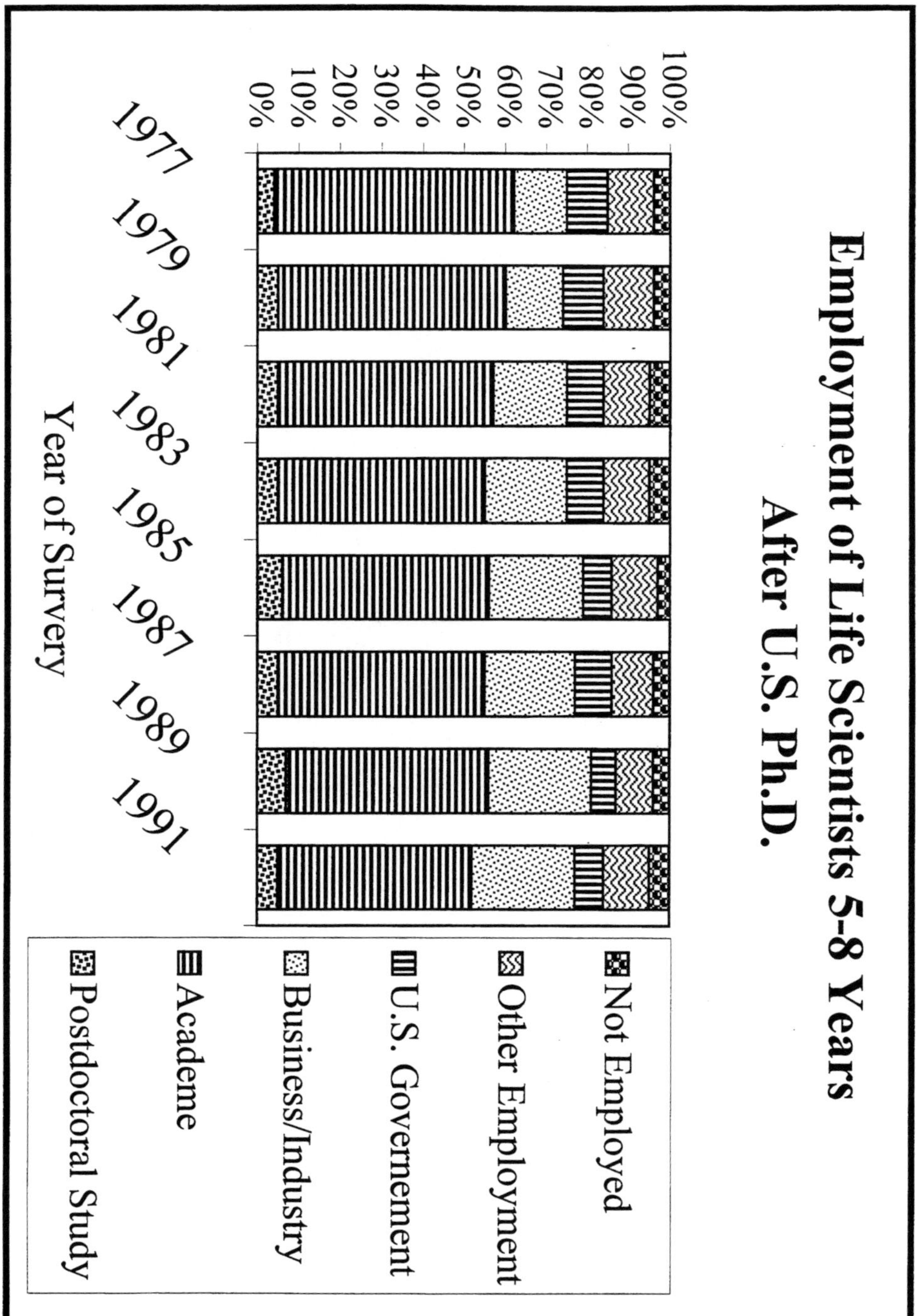
Employment of Life Scientists 5-8 Years
After U.S. Ph.D.
100%
90%
80%
70%
60%
50%
40%
30%
20%
10%
0%
1977
1979
1981
1983
1985
1987
1989
1991
Year of Survery
Not Employed
Other Employment
U.S. Governement
Business/Industry
Academe
Postdoctoral Study

Figure 2

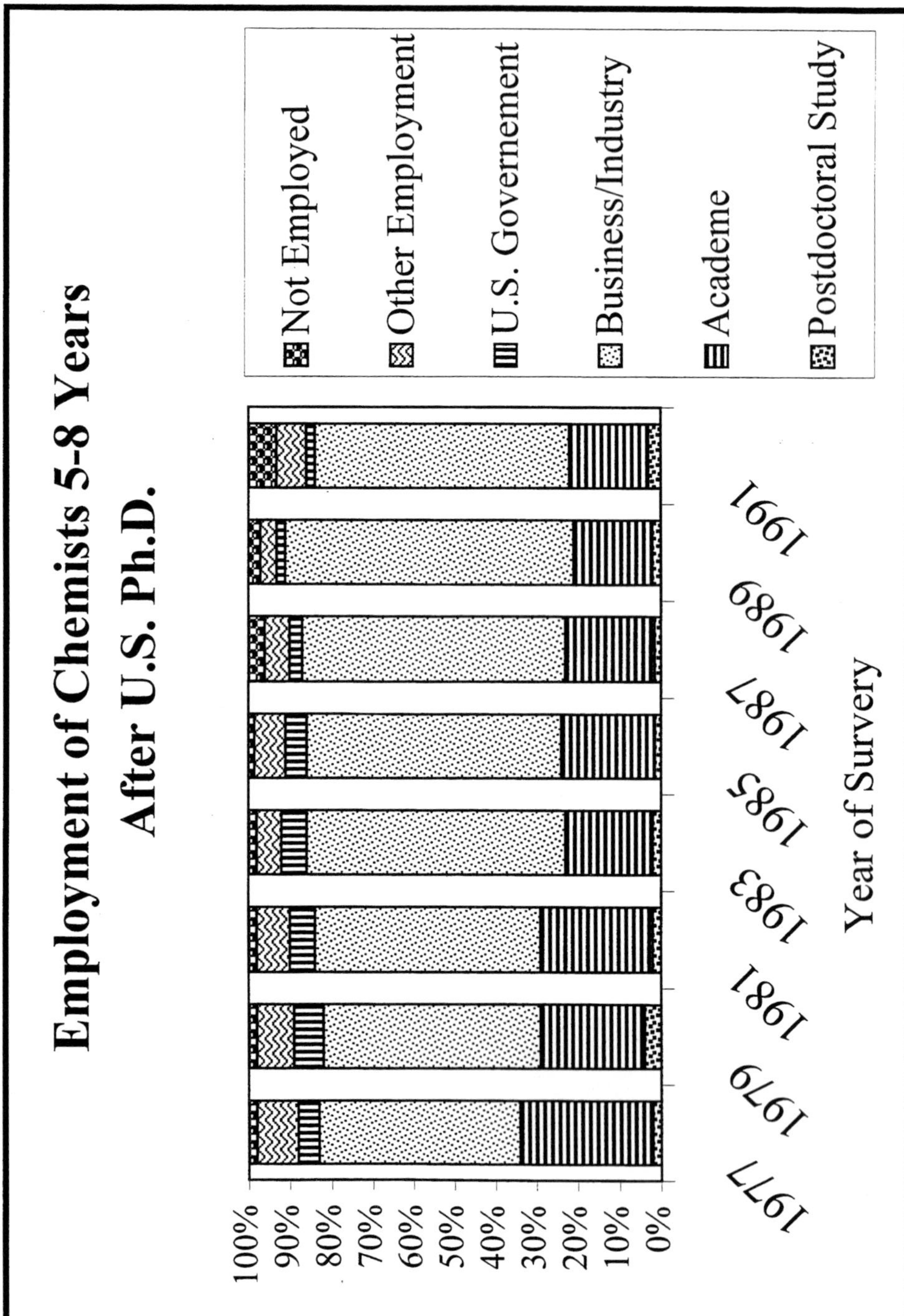
Employment of Chemists 5-8 Years
After U.S. Ph.D.
Not Employed
Other Employment
U.S. Governement
Business/Industry
Academe
Postdoctoral Study
100%
90%
80%
70%
60%
50%
40%
30%
20%
10%
0%
1977
1979
1981
1983
1985
1987
1989
1991
Year of Survery

in the life sciences in the 125 Carnegie Research I/II and 150 Doctoral-granting I/II institutions (my own institution is among these) from the nation's 2,641 institutions of higher learning.

There are several factors driving the increasing movement of graduate-trained life scientists into the corporate arena. There are currently 1,287 biotechnology companies in the nation with $1.5 billion in revenues (interestingly, making the entire biotechnology industry about 1.5 times the size of Merck which represent $100 billion in market capitalization). Currently, the entire industry has ~70 approved biological therapeutics in production, with about 40% annual increases in the number of new biologics introduced in the years of 1994-95 and 1995-96. Of significance, there are over 700 biologics in pre-clinical and clinical testing at the moment and of these, greater than 200 are in Phase 2 and 3. It is anticipated that about 30% of these will achieve FDA-approval.

The break-out of Phase 3 proto-products will have a substantial impact on the largely research-boutique, R&D-focused companies, moving them into the manufacturing and production arena with consequent impacts on the numbers of both non-research graduate and undergraduate trained employees (about 55% of pilot plant and 45% of manufacturing/production personnel are BS or higher-degreed). Naturally, movement into the profit-generating production mode will significantly enhance the basic research activities of the successful companies as they strive to generate new profitable and socially-responsive biotech products, resulting in greater needs for university-trained corporate research scientists.

From a conservative viewpoint, the use of chemical industry employment trends and needs as a basis for extrapolation for the biotechnology industry might be considered by some to be questionable. However, there are interesting projections on the extent to which biotechnological innovations in the non-biomedical area (the so-called second wave of biotechnology) will replace processes and technologies currently in the realm of the chemical sciences. Importantly, there are several wild cards which can have synergistic impacts on the growth of the biotechnology industry and employment trends, including the various genome projects that are currently underway. These knowledge and opportunity/business expanding genomic and related activities could dramatically amplify biotechnology workforce needs.

It is useful to track the real and projected workforce needs of the biotechnology industry over the past few years. The data in Figure 3 show the existing workforce for 1991-1996 compiled by Ernst and Young. At each year surveyed, the companies were queried regarding their expectations by the year 2000, and these data are seen in the upper curve (see Figure 3). Several years manifested over a 15% increase. In 1994 and 1995, growth was at 4% and 6%, respectively, largely due to the prominent number of mergers and acquisitions. Ernst and Young data for 1996 show a 9% increase. Currently, there is projected a biotechnology workforce of 250,000 to 1,000,000 by the year 2010, requiring an average annual growth of 5% or 17% annually, respectively. Figure 4 shows the region of uncertainty regarding the size of the biotechnology workforce of 2010.

I should emphasize that this workforce represents only the biotechnology industry, not the pharmaceutical industry. The latter represents a 264,000 employee, $20 billion profits, and $400 billion market capitalization industry. There are 71 members of the Pharmaceutical Research and Manufacturers Association, and 35 have a substantial biotechnology activity. It is expected that the non-biotechnology component of the pharmaceutical industry technical employee base will remain stable relative to their expanding biotechnology component, and these latter projected increases are not incorporated in Figures 3 and 4.

The technical workforce of a biotechnology company can be as high as 70%. Amongst this technical workforce, depending upon the survey and region, 11-13% are high school/community college trained, and 50-54% are BS, 16-17% are masters, and 19-20% are Ph.D. holders. I have had discussions with a number of my colleagues from several campuses who are considering the expansion of their masters level training efforts as a means of

Figure 3

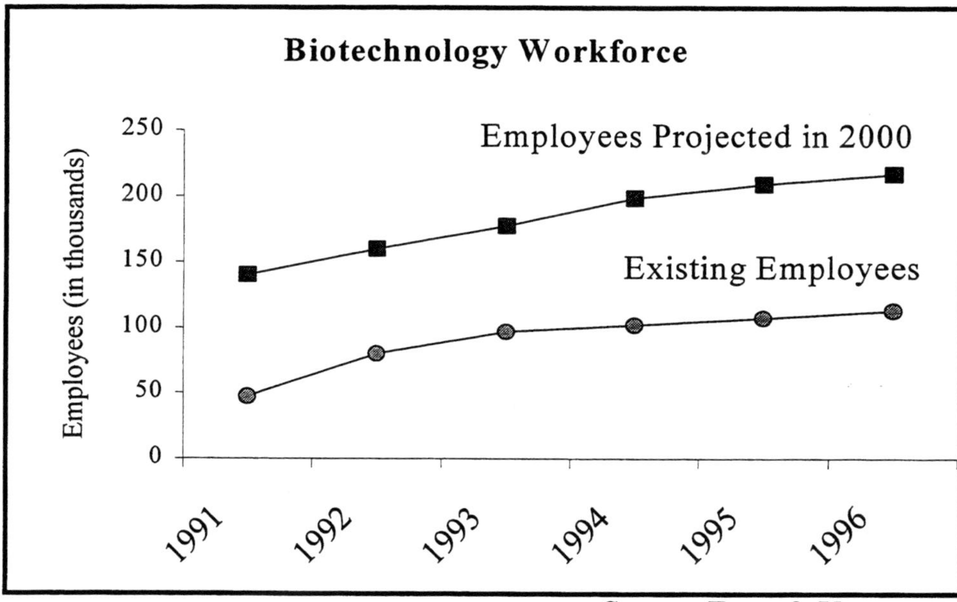

Source: Ernst & Young

Figure 4

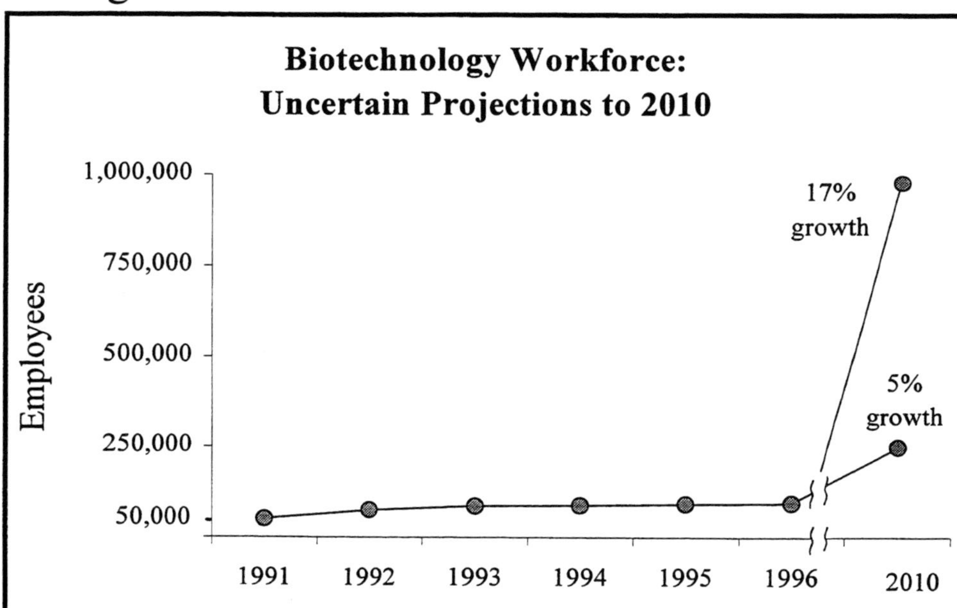

retaining stability in the research, research-training, and educational arenas as the numbers of needed Ph.D.s might be decreasing. It should be emphasized that two just-completed national surveys (one a 1993-95 retrospective of permanent corporate employees and the other a 1996-98 projective of future corporate hiring) have resulted in a nearly identical masters/Ph.D. ratio (1:1.17 vs. 1:1.18) for the molecular life science/biotechnology area.

Currently, the molecular life science/chemistry masters degree in the California State University (CSU) system is a research-based degree, requiring 2-3.5 years, whereas in UC biology and chemistry departments the masters degree might be awarded more frequently with a lesser degree of immersion in research or course work. Currently, it is fair to say that masters student training is not a major focus of the average UC faculty member. I have some strong concerns regarding UC expansion of masters programs without a vigorous research focus unless there is a clear delineation of the research masters vs. the non-research masters. Complicating the matter, the industry has identified areas where non-wet bench research, professional masters degrees are highly desirable. Many of my CSU colleagues tend to view such non-research based masters programs with a jaundiced eye, viewing such "parascientific" matters as detracting from their research and productive scholarship, but clearly there is a need for both university systems to consider creating new, tailored, non-research, professional masters programs.

The industry has made it clear that there are areas that universities and colleges can focus on which would produce a more valuable, contributory industry employee, without jeopardizing higher education's role and without invoking undesirable "vocationalistic trends". Clearly, those institutions that heed industry recommendations within the realm of the university philosophy and traditional modus operandi will be more competitive in contributing to the biotechnology workforce pool. One of these industry recommendations concerns the needs for more team-based activities within the university training environment, a proposal which seems contrary to the traditional university goal of developing independent student scientists. Interestingly, however, in response to industry needs, MIT is currently conducting experimental Ph.D. programs with up to eight Ph.D. candidates working on a tightly-focused, team-based project.

There are other aspects of graduate education that are undergoing change in response to industry and technological opportunities, including more management of technology foci and expansion of process-oriented training programs. There is also clearly a movement towards the use of distributed technologies and shared instruction on the graduate level. There are some high level discussions between the Universities of Massachusetts, Maryland, Minnesota and California regarding the use of distributive educational technologies in some of these parascientific arenas that are critically important for the industry.

Importantly, under normal conditions, introducing change into a graduate curriculum is not an easy process, and changes as we have discussed today require a not insignificant cultural and paradigm shift for many university faculty. I have spent a lot of time in the past few years educating faculty on what goes on in industry. Without a doubt they initially see the industry largely from their research perspective as being simply an admixture of cell culture and gene jockeying. In most cases they have no idea what goes on in the industry and that there are products involved and that the industry exists in a world of extreme constraints, rules, regulation, GMP, and a host of other acronyms for important concerns that are not in the lexicon of the average faculty member. In order to succeed in the expansion and responsiveness of our university programs, existing faculty have to be brought on board, minimally in validation that these endeavors are important.

I think clearly in California we will see an expansion of industry-responsive training platforms, ones requiring new resources. It is all together not clear that the State of California taxpayer wants to provide for the 170,000 students heading to CSU and another 60,000 into the University of California over the next eight to nine years. In order for these industry-responsive curricular and training changes to be carried out effectively, a very potent interaction between the academic community and the corporate area is needed in order to shake loose new resources from the State Legislature, and that's where organizations like the California Healthcare Institute and organizations like the University of California's Systemwide Biotechnology Program and CSU's Biotechnology Program own become very important in the sense of coordinating an aggressive and effective lobby for new legislative initiatives, as well as acquisition of both new public and private resources.

17 Charting a Course for the Future

M.R.C. Greenwood

Bruce Alberts and I share something in common. Someone -- in fact I think it was one of Bruce's graduate students -- has said to me, "Gee, Bruce has had an epiphany. He sees the world differently now that he's been to Washington." I will share a little story with you. It is one of my favorite stories and it will preface why I think I am seeing this particular discussion and perhaps the future differently than I might have some years ago.

The story started out as a joke that middle-aged men tell. I tell it now as a middle-aged woman. This nice middle-aged woman was walking through the woods one day when she heard a little voice saying, "Kiss me, kiss me!" She looked around, and as hard as she looked, she saw nothing and kept walking. Once again, the little voice said, "Kiss me, kiss me!" and the woman looked down and there was a very charming little frog sitting in the middle of the walkway. She picked it up and it said, "Kiss me. I am a handsome prince and if you'll just kiss me, I'll be yours." So she put the frog in her pocket and kept walking. The very indignant little frog said, "Didn't you hear what I said? I am a handsome prince. If you'll just kiss me, I'll be yours." And she said, "Oh, I heard you, but at this stage in my life I'd prefer a talking frog."

When I was a young girl my mother taught me that you have to kiss a lot of frogs before you find a prince, but I want to try to put today's discussion in a slightly different perspective. I look around and see many folks like me. Unfortunately, the discussion we are having today probably needs to be had with a larger group of

---

M.R.C. Greenwood is Chancellor of the University of California, Santa Cruz. This article is based on remarks delivered at the October 2, 1996 symposium, "Meeting the Challenge: Graduate Education in the Biological Sciences in the 21st Century" held at the University of California, San Francisco.

people: those who are facing the future and who will, in fact, be either the benefactors of our good intelligent choices for the future, or the victims of our less than intelligent choices. One of the things that I thought was very interesting about the convocation for graduate education that the National Academy pulled together in late spring of this year, was that it did include teams of folks from various institutions. Thus it included a large enough group of students that we were able to get some fresh voices and some different perspectives into discussions around the table, facilitating precisely what one of the speakers today said -- the kind of dialogue that is necessary for us to jointly face the future.

Among the many things that I think we must talk about as we attempt to chart the future for graduate education is the fact that very little has been said that actually bears on preparing the workforce for the 21st century. Are we really taking on the education responsibility that we bear? Are we looking at the options that we can provide to ensure that students are suitably prepared and continually renewed as they undertake some of the jobs that we believe will be important and interesting, and that will provide meaningful careers?

Thinking about Ph.D. education in the life sciences, five years ago I would have said that getting a Ph.D. could be described a little bit like laying a map of the United States out on the table and trying to find a way to get from San Francisco to New York. There are lots of ways to make the journey. One way to arrive as fast as possible is by car by getting on Interstate 80. Another, of course, is to get on an airplane and fly -- which is not always more direct. Another option is to choose a more variable route to see a good deal more of the country and still get to New York. These are the things that I think we have struggled with over the years. How long should graduate education be? How varied? How rich? Is it individual experience or is it intended to serve the needs of the faculty mentor?

We have found ourselves in recent times in the situation of pitying the concerns of the individual -- frequently the faculty member. This view contrasts with the more traditional and, in my time, more common way of looking at graduate education which was that, as someone has paraphrased recently, it takes a village to raise a graduate student. This is the more collective approach: a program which trains graduate students across a variety of subjects and doesn't tie them so closely to a single mentor or another individual on whom they may become dependent. We don't always appreciate that for our graduate students' future well-being, it matters not just whether they get their degrees, but how thoughtful and connected their mentors were in helping to enlarge the circle and network that they might have -- thereby increasing opportunities to understand their professional options. We have been without a systematic program within our graduate programs to help students understand the kinds of options available to someone who gets an advanced degree in the biomedical sciences.

Today, the problem is that we no longer have a two-dimensional map. We don't have the circumstances where we are simply trying to get from A to B. What is happening is that the world of the research university and the nature of work in the United States and around the world are changing very quickly. We are in a much more three-dimensional space. The impact of degree-getting is not so much a linear effect as it is what some folks have referred to as a sheepskin effect. When you get a particular degree, you get another bump up, so to speak. We know that for individuals in our society, every year of education beyond high school is directly correlated to their economic outcome. More education is a good thing -- and in a highly technical 21st century, more education ought to be a better thing. The statistics indicate that there is, in fact, an incremental difference for the years of education in technical fields.

I think this poses an interesting question for us. As the nature of work in the United States is changing and as universities themselves are changing, are we finding ourselves ever more in an ecosystem model of education as opposed to a linear model of degree production? I want to try to conceptualize this situation and also note that there are some very interesting books on the way businesses are organized that are relevant to this issue. One of the concepts that has been advanced is that we may want to characterize the 21st century as the age of imagination as opposed to the Information Age, an outdated concept. We are talking about a century in which knowledge -- the ability to use knowledge, and the ability to use the tools that allow you to acquire knowledge -- are the job requirements for the future.

The article by Edward Penhoet (Chapter 5) describes the death of the traditional job. Businesses are, themselves, no longer going to be so monolithic and so linear. They frequently are going to be smaller and very different. James Moore, in "The Death of Competition", writes that we are going to have businesses which compete on one day, collaborate on another day, and work in consortiums on yet another. That is probably how higher education ought to be working over the next 25 years -- in a mode where we are actually working together, not just in consorta of graduate programs but in consorta of institutions that enable us to train students in a variety of ways. The big question is: are we charting a course for the future? Are we going to define the Ph.D. degree as a very specialized degree that has a couple of purposes (similar to that described in the article by Keith Yamamoto, Chapter 15)? Are we going to develop a well-trained mind who has done some significant work and is capable of going on to become an independent investigator in industry or academia? A big piece missing from this is that nothing was said about training the individual, not just as independent investigators, but to function in today's and the future university.

I would argue that universities will still be large employers of Ph.D.s, although they will certainly not be the only ones. Little is said about inculcating our students with the capacity to teach or the ability to understand the dynamics of the classroom. There is little recognition that universities will be hiring faculty members who not only are acquainted with high technology, but who are, in fact, very good at multi-media presentations. Such faculty, who understand the power of a cognitive discipline and the knowledge that they have developed, must form pedagogical techniques for the future. At least, we have to accept the possibility that not everybody will forever be taught as they are today.

Some of the people we are training at the advanced level are going to have to be leaders in integrating the cognitive knowledge about how people learn in the classroom and other settings. If we are not successful at training the next generation of professors and academic leaders, there won't be a level of public support for the missions of the university within the State of California. If the public doesn't trust us to educate well, if they don't believe that we know how to teach, and if they don't believe that we are better than Disney and Intel, then why would they give us their business? Outcomes assessment, and teaching pedagogy that allows you to assess whether or not your students have learned, are both essential in advancing these fields. At the same time, they are advancing our understanding of our audience and our communication skills.

Frankly, people with these skills are not commonly found in some of our research universities. We stand a real chance that the careers for which we think we are training our Ph.D.s -- in particular, to be professors like us -- are changing fast enough that if we don't make some changes in our training programs, we won't fill them. It will not just be that we do not know how to teach them how to work in tomorrow's university. That is a challenge about which I have heard very few intelligent discussions, anywhere.

What are the jobs of the future going to be for people who have doctoral degrees? How are we going to prepare them for these jobs, beyond training their scientific minds and developing their research skills? Maybe that is enough. But if it is enough, we should say that it is enough. We should decide that students do not have to achieve Nobel Prize-quality research to leave the university with a Ph.D. They don't need to stay eight or ten years. They need to finish more rapidly and with a much higher quality of generalist training. They need to have a much higher level of confidence in their own ability to master fields independently and to develop other skills necessary for their future careers.

Edward Penhoet (Chapter 5) discusses the fact that general training is actually more adaptable than highly specialized training. I agree with him that universities are going to have to recognize this, and we are already being called upon to do so. Universities are going to become knowledge centers that provide life long learning. Part of what we have to teach our graduate students is how to continue to access that knowledge and how to be comfortable with new demands. I thought it was very interesting that both Stanford and UC San Francisco are developing ways of admitting students to a general area of biological sciences with the intent that once the students are there, they will figure out which of the people they want to study with and in which of the specializations they would like to focus their degree. In fact, that is not a new model. I am a graduate of Rockefeller University and that has been the model there and, as far as I know, it still is. All students are admitted to the university, not to discrete graduate programs. Within the university they're allowed to move around. You can come in as a biologist and leave as a philosopher, or at least that was possible when I was a student. Perhaps you can't leave as a philosopher anymore, but you can leave as a mathematician or as a chemist, instead of as a biologist.

As we chart the course for the future, we need to worry not so much about the course in the traditional sense of a certain future and concrete steps that we know can guide our students to valuable careers. We must ask about the real value to be gained by completing a Ph.D. We must ask what are the skills we really expect people to have when they leave with that Ph.D. What documentation is considered satisfactory to validate those skills? Then, how can we continue to stay in touch with those students as they go forward in their careers? How can institutions of higher learning provide nodes for the Age of Imagination and continuing access to the kind of knowledge and even periodic retooling that people will need to adapt to changes -- in particular to changes in the type of role that a faculty member leads 25 years from now in a research university.

I urge us to think, not just what other jobs do we currently know about, but also appear in the future. We might be giving our graduate students broadening experiences and exposures to other opportunities. The Committee on Science, Engineering and Public Policy (COSEPUP) report focused on the fact that there are significant opportunities out there that our students don't know about. I have noticed, just from my own perusal of the Web, that there has been a great response to that report. There now is a lot of information that students can access, including career guides and job postings all over the World Wide Web that are much more timely than printed advertisements. We have begun to respond to the issues raised in the COSEPUP report. Faculty and administrators have started to talk to graduate students about careers other than exactly the one that I have and that you have. What we haven't talked about, though, is whether even the expectations might be changing of what institutions would be hiring faculty members to do in the next 25 years.

In order to chart the course -- or better, to help our students chart their own courses -- think we are going to have to talk about that side of the equation, as well. Finally, if we are going to talk about training in the biological sciences in the 21st century, we will have to meet the challenge of just-in-time training. It is a concept that is alien to most of my faculty colleagues, but one that is quite familiar to our business colleagues. The concept is: how much training, of what kind, and at what time in your career.

Most of the discussion here has focused on research at Research I universities. We must remind ourselves that research universities will train less than 20% of the national workforce of the 21st century. Thus, we need some concourse and discussion among the other segments of higher education that are, in fact, producing the large number of Ph.D.s and Master students in the country.

I would like to be able to say that I have a clear vision of what graduate education for the next 25 years should be and that I have a chart for that future. If I had that chart, I'd probably be able to go back to Washington and sell something new. But I don't have it. I agree with the article by Richard Attiyeh (Chapter 11) that the chart is going to be drawn by collective consciousness raising (to go back to a good old '60's term) of both the groups that are involved in training students and the students, themselves. We have gotten to an advanced stage in education specialization that makes it very difficult for a student to arrive at an intelligent choice about where they are going to go in the biological sciences. I applaud efforts to broaden the way in which students enter and pursue advanced degree programs in universities, and efforts to give them the option to make decisions after they have a broader understanding of the field. This makes sense.

Students are making choices in which they are going to have to invest (even with the model described in the article by Keith Yamamoto, Chapter 15) probably eight to nine years to become independent investigators in industry or in academia. They are making lifelong commitments in their junior year -- after they have had at most four or five courses in the discipline. We should give them greater opportunity to graze in the field before we ask them to choose their favorite grass.

Meeting the Challenge: Graduate Education in the Biological Sciences in the 21st Century
Wednesday, October 2, 1996
University of California, San Francisco, Laurel Heights Campus

Participants

Dr. Tony Adams
Dean
School of Optometry
UC, Berkeley
Berkeley, California 94720
Ph: 510-642-3414

The Hon. Alfred E. Alquist
Senator
District 13 (D-San Jose)
State Capitol
California State Senate
P.O. Box 942848
Sacramento, California 94248
Ph: 916-445-9740

Elaine White Alquist
State Assembly Candidate
105 Town & Country Village
Sunnyvale, California 94086
Ph: 408-739-1996

Mr. Steve Arthur
Assistant Director
California Trade
and Commerce Agency
801 K St., Suite 1700
Sacramento, California 95814
Ph: 916-322-3524

Dr. Clifford Attkisson
Dean of Graduate Studies
Graduate Division
200 Millberry Union West
Box 0244
500 Parnassus Avenue
UC, San Francisco
San Francisco, California 94143
Ph: 415-476-2310

Dr. Dorothy F. Blainton
Vice Chancellor,
Academic Affairs and Professor of Pathology
Medical Sciences Building,
Box 0400
UC, San Francisco
San Francisco, California 94143
Ph: 415-476-4451

Dr. Colin Bloor
Professor of Pathology
Director, Molecular Pathology
Graduate Program
School of Medicine
9500 Gilman Drive
UC, San Diego
La Jolla, California 92093
Ph: 619-534-4324
E-mail@ cbloor@ucsd.edu

William Bold
Director of Public Policy
California Health Care Institute
1020 Prospect St., Suite 310
La Jolla, California 92037
Ph: 619-551-6677

Dr. Barry Bowman
Professor
Department of Biology
225 Sinsheimer Labs
UC, Santa Cruz
Santa Cruz, California 95064
Ph: 408-459-4120
E-mail: bowman@
biology.ucsc.edu

Mr. Jess Bravin
UC Regent
Boalt Hall
UC, Berkeley
Berkeley, California 94720
Ph: 510-642-5081
E-mail: jessb
@uclink2.berkeley.edu

Dr. George Bruening
Director
Department of Plant Pathology
559 Hutchison Hall
1930 Fifth St.
UC, Davis
Davis, California 95616
Ph: 916-752-3474
E-mail: bgbruening
@ucdavis.edu

Dr. Alma Burlingame
Professor
School of Pharmacy
S 1124, Box 0446
Department
of Pharmaceutical Chemistry
513 Parnassus Ave.
UC, San Francisco
San Francisco, California 94143
Ph: 415-476-5641
E-mail: alb@itsa.ucsf.edu

Dr. Ken Burtis
Associate Professor & Chair
Genetics Graduate Group
Department of Molecular
and Cellular Biology
UC, Davis
Davis, California 95616
Ph: 916-752-4188
E-mail: kcburtis@ucdavis.edu

Dr. Steven Cassedy
Chairman, Coordinating Council
on Graduate Affairs
Professor,
Department of Literature
9500 Gilman Dr.
UC, San Diego
La Jolla, California 92093
Ph: 619-534-3343
E-mail: scassedy@ucsd.edu

David Commons
UC Biotechnology Advisory
Committee
1835 Camino Palermo
Los Angeles, California 90046
Ph: 213-876-0124

Dr. Charles S. Craik
Professor
Pharmaceutical Chemistry
926 Medical Sciences Building
Box 0446
513 Parnassus Avenue
UC, San Francisco
San Francisco, California 94143
Ph: 415-476-8146
E-mail: craik@cgl.ucsf.edu

Dr. Linda De Young
Scientist
Genentech, Inc.
460 Point Bruno Blvd.
So. San Francisco, CA 94019
Ph: 415-225-1947
E-mail: linda@gene.com

Dr. Haile T. Debas
Dean
School of Medicine
S224 Medical Sciences Building
Box 0410
UC, San Francisco
San Francisco, California 94143
Ph: 415-476-2341
E-mail: hdebas
@medsch.ucsf.edu

Dr. Ken A. Dill
Professor
Pharmaceutical Chemistry
926-S/Pharmacy
UC, San Francisco
San Francisco, California 94143
Ph: 415-476-9964
E-mail: dill@maxwell.ucsf.edu

Mr. Fred Dorey
President
Bay Area Bioscience Center
1300 Clay Street, Suite 320
Oakland, California 94612
Ph: 510-874-1464

Mr. Orlando Elizondo
Legislative Policy Coordinator
103 Laurel Heights, Box 0462
Community
and Governmental Relations
UC, San Francisco
San Francisco, California 94143
Ph: 4150476-8431

Dr. Anthony Fink
Professor, Department
of Chemistry and Biochemistry
389 Sinsheimer Labs
1156 High Street
UC, Santa Cruz
Santa Cruz, California 95064
Ph: 408-459-2744
E-mail: enzyme@cats.ucsc.edu

Dr. Robin S. Fisher
Associate Dean
Graduate Division
1237 Murphy Hall
405 South Hilgard Avenue
UC, Los Angeles
Los Angeles, California 90095
Ph: 310-825-0843
E-mail: rfisher@gdnet.ucla.edu

Dr. Sharon Fleming
Associate Dean of Research
College of Natural Resources
Dean's Office,
101 Giannini Hall
UC, Berkeley 94720
Ph: 510-642-7171
E-mail: fleming
@nature.berkeley.edu

Dr. Jean Fort
Assistant Dean
Graduate Studies and Research
ERC, 003
UC, San Diego
La Jolla, California 92093
Ph: 619-534-3683
E-mail: jfort@ucsd.edu

Dr. Joy S. Frank
Sr. Associate Dean for Research
School of Medicine
Room 12-138 CHS
10833 Le Conte Avenue
UC, Los Angeles
Los Angeles, California 90095
Ph: 310-825-8680
E-mail: joy
@deans.medsch.ucla.edu

Dr. Joyce B. Freedman
Director
Sponsored Project Office
336 Sproul Hall, 5940
UC, Berkeley
Berkeley, California 94720
Ph: 510-642-8110
E-mail: jbfreed
@uclink2.berkeley.edu

Dr. Edward Furtek
Associate Vice Chancellor
Torrey Pines Center North
Office of Science
and Technology Policy
and Programs
10300 N. Torrey Pines Road
UC, San Diego
La Jolla, California 92093
Ph: 619-534-8400
E-mail: efurtek@ucsd.edu

Brenda Gee
Director
Admissions and Student
Academic Progression
Graduate Division, 0404
513 Parnassus Avenue, S-140
UC, San Francisco
San Francisco, California 94143
Ph: 415-502-4460
E-mail: grenda_gee_atsaa@
ccmail.ucsf.edu

Dr. Laural Gershwin
Professor
Department of Pathology,
Microbiology, & Immunology
School of Veterinary Medicine
UC, Davis
Davis, California 95616
Ph: 916-752-6643
E-mail: ljgershwin@ucdavis.edu

Dr. Jeffery Gibeling
Executive Associate Dean
Graduate Studies
252 Mrak Hall
UC, Davis
Davis, California 95616
Ph: 916-752-0652
E-mail: jcgibeling@ucdavis.edu

Dr. Joe W. Gray
Department of Cancer Genetics
Cancer Center,
MCB 230, Box 0808
UC, San Francisco
San Francisco, California 94143
Ph: 415-476-3461
E-mail: gray@dmc.ucsf.edu

Dr. Harry Green
Vice Chancellor for Research
200 University Office Building
UC, Riverside
Riverside, California 92521
Ph: 909-787-5535
E-mail: harry.green@ucr.edu

Dr. Victoria Hale
Scientist
Pharmacological Science
Genentech, Inc.
460 Point San Bruno Boulevard
So. San Francisco, CA 94080
Ph: 415-225-8386
E-mail: vhale@gene.com

Dr. Barbara Hamkalo
Professor and Acting Dean
Dept. of Molecular Biology
122 BSA
UC, Irvine
Irvine, California 92697
Ph: 714-824-7168
E-mail: bahamkal@uci.edu

Dr. Christina K. Hansen
Director, Contracts and Grants
Research and Graduate Studies
155 Administration Building
UC, Irvine
Irvine, California 92769
Ph: 714-824-5677
E-mail: ckhansen@uci.edu

Paul Hanson
Graduate Student
Department of Chemistry
and Biochemistry
UC, Santa Cruz
Santa Cruz, California 95064
Ph: 408-459-3390
E-mail: hanson
@hydrongen.ucsc.edu

Dr. Ronald W. Henderson
Dean
Graduate Studies and Research
399 Applied Science Building
UC, Santa Cruz
Santa Cruz, California
Ph: 408-459-2510
E-mail: henders@cats.ucsc.edu

Dr. Harvey Roy Herschman
Professor
Department of Biological Chemistry/Pharmacology
341 Molecular Biology Institute
611 Circle Drive East
UC, Los Angeles
Los Angeles, California 90095
Ph: 310-825-8735
E-mail: harvey@lbes.medsch.ucla.edu

Dr. John Hershey
Professor
Dept. of Biological Chemistry
UC, Davis
Davis, California 95616
Ph: 916-752-3516
E-mail: jwhershey@ucdavis.edu

Dr. Karl J. Hittleman
Associate Vice Chancellor
Academic Affairs
Research Chancellor
115 Medical Sciences Building, Box 0400
UC, San Francisco
San Francisco, California 94143
Ph: 415-476-1883
E-mail: khittlem@chanoff.ucsf.edu

Dr. C. Anthony Hunt
Chairman
Dept. of Pharm. Chemistry
School of Pharmacy
UC, San Francisco
San Francisco, California 94143
Ph: 415-476-2455
E-mail: hunt@itsa.ucsf.edu

Dr. Thomas James
Professor and Chairman
Pharmaceutical Chemistry
HS-1109, Box 0446
513 Parnassus Avenue
UC, San Francisco
San Francisco, California 94143
Ph: 415-476-1916
E-mail: james@picasso.ucsf.edu

Dr. Bing K. Jap
Senior Staff Scientist
Donner Laboratory, Room 363
Lawrence Berkeley National Lab
Berkeley, California 94720
Ph: 510-486-7990
E-mail: jap@rhoda.lbl.gov

Joyce B. Justus
Special Assistant to the Provost
Academic Affairs
Kaiser Building 18th Floor
UC Office of the President
300 Lakeside Drive
Oakland, California 94612
Ph: 510-987-9413
E-mail: joyce.justus@ucop.edu

Dr. Stephen B. Kahl
Professor
Chemistry and Pharm. Chem.
S-1122
Department of Pharm. Chem.
513 Parnassus Avenue
UC, San Francisco
San Francisco, California 94143
Ph: 415-476-4684
E-mail: sbkahl@itsa.ucsf.edu

Dr. Harumi Kasamatsu
Professor
Dept. of Molecular, Cell,
and Developmental Biology
405 Hilgard Avenue
UC, Los Angeles
Los Angeles, California 90095
Ph: 310-825-3048

Dr. Noel Keen
Professor
Department of Plant Pathology
2401 Boyce Hall
UC, Riverside
Riverside, California 92521
Ph: 909-787-4134
E-mail: keen@ucrac1.ucr.edu

Dr. Regis Kelly
Professor and Chair
Department of Biochemistry
and Biophysics
UC, San Francisco
San Francisco, California 94143
Ph: 415-476-4095

Dr. George Kenyon
Dean and Professor
School of Pharmacy
S-926, Box 0446
513 Parnassus Avenue
UC, San Francisco
San Francisco, California 94143
Ph: 415-476-8010
E-mail: glk@itsa.ucsf.edu

Lisa Kim-Shapiro
Graduate Student
Dept. of Pharm. Chemistry
Box 0446
UC, San Francisco
San Francisco, California 94143
Ph: 415-476-5674
E-mail: jwkim@cgl.ucsf.edu

John D. Kluge
Staff Research Associate
Graduate School of Management
County Rd. 98 & Hutchison Rd.
UC, Davis
Davis, California 95616
Ph: 916-752-0447
E-mail: jdkluge@ucdavis.edu

Dr. Mary Anne Koda-Kimble
Professor and Chair
Clinical Pharmacy
C-152, Box 0622
521 Parnassus Avenue
UC, San Francisco
San Francisco, California 94143
Ph: 415-476-2354
E-mail: makx@itsa.ucsf.edu

Dr. Karen Kovacs
Special Asst. to the Vice
Chancellor for Academic Affairs
1237 Murphy Hall
405 South Hilgard Avenue
UC, Los Angeles
Los Angeles, California 90095
Ph: 310-825-4427

Dr. Rosemarie H. Kraft
Associate Dean
Graduate Studies
252 Mrak Hall
UC, Davis
Davis, California 95616
Ph: 916-752-0652
E-mail: rhkraft@ucdavis.edu

Dr. Mary Kreger
Senior Analyst
Special Research Programs
UC Office of the President
300 Lakeside Drive
Oakland, California 94612
Ph: 510-987-9885

Lisa Krieger
Medical Reporter
San Francisco Examiner
110 Fifth Street
San Francisco, California 94103
Ph: 415-777-7867

Dr. Charlotte Kuh
Executive Director
Office of Scientific
and Engineering Personnel
National Research
Council/National Academy
of Sciences
2101 Constitution Avenue, N.W.
Washington, D.C., 20418
Ph: 202-334-2700
E-mail: ckuh@nas.edu

Dr. Irwin Kuntz
Professor
Department of Pharm. Chemistry
513 Parnassus Avenue
UC, San Francisco
San Francisco, California 94143
Ph: 415-476-1937
E-mail: kuntz@cgl.ucsf.edu

Dr. Arthur D. Lander
Associate Professor
Department of Developmental and Cell Biology
5205A Biological Sciences II
UC, Irvine
Irvine, California 92697
Ph: 714-824-1008

The Honorable Barbara Lee
Assembly Member
District 16 (D-Oakland)
State Capitol
California State Assembly
P.O. Box 942849
Sacramento, California 94249
Ph: 916-445-7442

Dr. Frances Leslie
Associate Vice Chancellor for Research
Office of the Vice Chancellor
155 Administration
UC, Irvine
Irvine, California 92697
Ph: 714-824-2579
E-mail: fmleslie@uci.edu

Dr. Barbara Levey
Assistant Vice Chancellor
Biomedical Affairs
2107D Murphy, Box 951405
Chancellor Office
UC, Los Angeles
Los Angeles, California 90095
Ph: 310-206-4656
E-mail: blevey@conet.ucla.edu

Dr. Charles N. Li
Dean
Graduate Division
3117 Cheadle Hall
UC, Santa Barbara
Santa Barbara, California 93106
Ph: 805-893-4489

Dr. Paul Licht
Dean
Biological Sciences
201 Campbell Hall, 2920
UC, Berkeley
Berkeley, California 94701
Ph: 510-642-5716
Email:licht@garnet.berkeley.edu

Terry Lightfoot
Public Info Representative
News and Communications
UC Office of the President
300 Lakeside Drive
Oakland, California 94612
Ph: 510-987-9194

Dr. Robert A. Ludwig
Dept. of Biochemistry
and Molecular Biology
425 Sinsheimer Laboratories
UC, Santa Cruz
Santa Cruz, California 95064
Ph: 408-459-4084
E-mail: ludwig@orchid.ucsc.edu

Dr. Roger Lundblad
Director
Science and Technology Dev't
Baxter Biotech, Hyland Division
1720 Flower Avenue
Duarte, California 91010
Ph: 818-305-6033
E-mail: lundblr@baxter.com

Rick Malaspina
Associate Director
for News Services
News and Communications
UC Office of the President
300 Lakeside Drive
Oakland, California 94612
Ph: 510-987-9179

Dr. Jerry Manning
Professor and Chair
Molecular Biology and Biochem
2242 Biological Sciences II
UC, Irvine
Irvine, California 92697
Ph: 714-824-5578
E-mail: jemannin@uci.edu

Dr. Joseph Martin
Chancellor
126 Medical Sciences Building
513 Parnassus Avenue
UC, San Francisco
San Francisco, California 94143
Ph: 415-476-2401
Email:joemart@chanoff.ucsf.edu

Dr. Niall Mateer
Director, Research Outreach
and Initiatives
UC Office of the President
300 Lakeside Drive, 18th Floor
Oakland, California 94612
Ph: 510-987-9478
E-mail: nial.mateer@ucop.edu

Dr. Carol McClain
Director
Multicampus Research
and Planning Programs
UC Office of the President
300 Lakeside Drive, 18th Floor
Oakland, California 94612
Ph: 510-987-9473
E-mail: carol.mcclain@ucop.edu

Dr. David McDonald
Assistant Dean
for Student Affairs
Office of Graduate Studies
and Research
9500 Gilman Drive
UC, San Diego
La Jolla, California 92093
Ph: 619-534-3871
E-mail: dmcdonald@ucsd.edu

Dr. Martina McGlouhlin
Associate Director
Biotechnology Program
1201 Meyer Hall, 8689
UC, Davis
Davis, California 95616
Ph: 916-752-3260
E-mail: mmmcgloughlin
@ucdavis.edu

Dr. Mark G. McNamee
Dean
Division of Biological Sciences
355 Briggs Hall
UC, Davis
Davis, California 95616
Ph: 916-752-6764
Email:mgmcname@ucdavis.edu

Dr. Karen Merritt
Director, Academic Planning
and Program Review
UC Office of the President
300 Lakeside Drive, 18th Floor
Oakland, California 94612
Ph: 510-987-9468
E-mail: Karen.Merritt@ucop.edu

Dr. Susan M. Miller
Assistant Professor
Pharmaceutical Chemistry
1117A S, Box 0446
513 Parnassus Avenue
UC, San Francisco
San Francisco, California 94143
Ph: 415-476-7155
E-mail: smiller@cgl.ucsf.edu

Dee Minor
Staff-Registrar
345 Giannini Hall, #3100
UC Systemwide Biotech Progrm
UC, Berkeley
Berkeley, California 94720
Ph: 510-643-0725

Paul Monroe
Brigadier General
California National Guard
1892 San Juan Avenue
Berkeley, California 94707
Ph: 510-526-9331
E-mail: pdmon2@aol.com

Judi Mozesson
Manager
Pharmaceutical Chemistry
UC, San Francisco
San Francisco, California 94143
Ph: 415-476-1915
E-mail: mozesson@cgl.ucsf.edu

Dr. Jim Murray
Professor
Department of Animal Science
UC, Davis
Davis, California 95616
Ph: 916-752-3179
E-mail: jdmurray@ucdavis.edu

Dr. Mary Napier
Director
Research Administration
Genentech, Inc.
460 Point San Bruno Boulevard
S. San Francisco, CA 94080
Ph: 415-225-1138
E-mail: napier.mary@gene.com

Jeremiah Paknawin-Mock
President
Graduate Student Association
MU112W
UC, San Francisco
San Francisco, California 94143
Ph: 415-476-2233
E-mail: jpmock@itsa.ucsf.edu

Linda Palmer
Staff-Registrar
345 Giannini Hall
UC Systemwide Biotech Prgrm
UC, Berkeley
Berkeley, California 94720
Ph: 510-643-0725

Dr. Bernhard O. Palsson
Professor
Department of Bioengineering
Engineering Bldg., Unit 1
9500 Gilman Drive
UC, San Diego
La Jolla, California 92093
Ph: 619-534-5668
E-mail: bpalsson
@bioeng.ucsd.edu

Dr. Rodney Pearlman
Vice President
Research and Development
Megabios Corporation
863A Mitten Road
Burlingame, California 94010
Ph: 415-697-1900 x220
E-mail: rodney@megabios.com

Patricia Pelfrey
Assistant to the President
UC Office of the President
300 Lakeside Drive, 22nd Floor
Oakland, California 94612
Ph: 510-987-9091
Email:patricia.pelfrey@ucop.edu

Dr. John Penhune
Senior Vice President, Research
Science Applications
International Corporation
1200 Prospect St., Suite 400
La Jolla, California 92037
Ph: 619-546-6810
E-mail: john_penhune
@cpqm.saic.com

Dr. Michael E. Phelps
Professor
School of Medicine
23-132 CHS, Box 951735, 1735
UC, Los Angeles
Los Angeles, California 90095
Ph: 310-825-6539
E-mail: mphelps
@mail.nuc.ucla.edu

Dr. Carl Poppe
Associate Vice Provost,
Research
and Laboratory Programs
UC Office of the President
300 Lakeside Drive, 18th Floor
Oakland, California 94612
Ph: 510-987-9405
E-mail: carl.poppe@ucop.edu

Dr. James Posakony
Professor
Department of Biology
9500 Gilman Drive
UC, San Diego
La Jolla, California 92093
Ph: 619-534-6300
E-mail: jposajony@ucsd.edu

Dr. Michael F. Powell
Senior Scientist
Group Leader
Genentech, Inc.
460 Point San Bruno Boulevard
S. San Francisco, CA 94080
Ph: 415-225-1389
E-mail: powell.mike@gene.com

Dr. Gordon C. Rausser
Dean, College of Nat. Resources
101 Giannini Hall
UC, Berkeley
Berkeley, California 94720
Ph: 510-642-7171
E-mail: rausser
@nature.berkeley.edu

Dr. Kenneth Raymond
Professor and Chair
Department of Chemistry
531 Latimer Hall, 1460
UC, Berkeley
Berkeley, California 94720
Ph: 510-642-7219
E-mail: raymond
@garnet.berkeley.edu

Dr. David Richman
Professor and Chair
Department of Neurology
Main Hospital, 5308
UC, Davis
Davis, California 95616
Ph: 916-754-5035
E-mail: dprichman@ucdavis.edu

Celeste E. Rose
Assistant VP
University Relations
UC Office of the President
300 Lakeside Drive, 22nd Floor
Oakland, California 94612
Ph: 510-987-9331
E-mail: celeste.rose@ucop.edu

Dr. Louise Rosenbaum
Assistant Director
UC Systemwide Biotech Prgrm
345 Giannini Hall
UC, Berkeley
Berkeley, California 94720
Ph: 510-643-0725

Dr. Michael K. Rust
Professor and Chairman
Department of Entomology, 134
UC, Riverside
Riverside, California 92521
Ph: 909-787-5327
E-mail: mikerust@mail.ucr.edu

Dr. Robert Sacher
Biotechnology Coordinator
Hunt-Wesson, Inc.
1645 W. Valencia Drive
Fullerton, California 92633
Ph: 714-680-2811

Dr. Charles Samuel
Professor and Chair
Department of Molecular,
Cellular, and Devel. Biology
4165 Biological Sciences II Bldg
UC, Santa Barbara
Santa Barbara, California 93106
Ph: 805-893-3097
E-mail: samuel@lscf.ucsb.edu

Dr. Barbara A. Sawrey
Vice Chair for Education
Department of Chemistry
and Biochemistry
9500 Gilman Drive
UC, San Diego
La Jolla, California 92093
Ph: 619-534-6479
E-mail: bsawrey@ucsd.edu

Dr. Howard Schachman
Professor
Dept. of Molecular
and Cell Biology
Stanley Hall
UC, Berkeley
Berkeley, California 94710
Ph: 510-642-7046
E-mail: schach
@garnet.berkeley.edu

Dr. Walter Shaffer
Director
Research Training
National Institutes of Health
6701 Rockledge Drive
Bethesda, Maryland 20982
Ph: 301-435-2687
E-mail: wsllq@nih.gov

Dr. Thomas Schleich
Professor and Chairperson
Department of Chemistry
and Biochemistry
Thimann Laboratories
UC, Santa Cruz
Santa Cruz, California 95064
Ph: 408-459-2067
E-mail: yoti@aku.ucsc.edu

Dr. Barbara O. Schneeman
Dean
Agricultural
and Environmental Sciences
275 Mrak Hall
UC, Davis
Davis, California 95616
Ph: 916-752-1605
E-mail: deancaes
@agdean.ucdavis.edu

Dr. David Shelby
Assistant Dean
Division of Biological Sciences
355 Briggs Hall
UC, Davis
Davis, California 95616
Ph: 916-752-3478
E-mail: drshelby@ucdavis.edu

Dr. Robert Shelton
Vice Chancellor for Research
Office of the Vice Chancellor
410 Mrak Hall, 8671
UC, Davis
Davis, California 95616
Ph: 916-752-4091
E-mail: rnshelton@ucdavis.edu

Sonephet Siharath
Staff-Registrar
345 Giannini Hall
UC Systemwide Biotech Prgrm
UC, Berkeley
Berkeley, California 94720
Ph: 510-643-0725

Dr. Lloyd Smith
Associate Dean
214 Medical Sciences Building
School of Medicine
513 Parnassus Avenue
UC, San Francisco
San Francisco, California 94143
Ph: 415-476-2342
Email: hsmith@medsch.ucsf.edu

The Honorable Jackie Speier
Assembly Member
District 19 (D-Burlingame)
State Capitol
California State Assembly
P.O. Box 942849
Room 4140
Sacramento, California 94249
Ph: 916-445-8020

Dr. Larry Stein
Professor and Chair
Department of Pharmacology
Medical Surge II, 360
College of Medicine
UC, Irvine
Irvine, California 92668
Ph: 714-824-6771

Dr. John Stephenson
Chief Operating Officer
Santa Cruz Biotechnology
2161 Deleware
Santa Cruz, California 95060
Ph: 408-457-3800

Dr. Jay Stowsky
Director, Research Policy
and Development
UC Office of the President
300 Lakeside Drive, 18th Floor
Oakland, California 94612
Ph: 510-987-9428
E-mail: jay.stowsky@ucop.edu

Dr. Michael P. Stryker
Professor and Chair
Department of Physiology
513 Parnassus Avenue, S-762
UC, San Francisco
San Francisco, California 94143
Ph: 415-476-8371
E-mail: stryker@phy.ucsf.edu

Judi Sui
Analyst, Graduate Division
427 Sproul Hall, 5900
UC, Berkeley
Berkeley, California 94720
Ph: 510-643-7478
Email: jsui@uclink.berkeley.edu

Dr. Ellen Switkes
Assistant to the Vice President
for Academic Advancement
UC Office of the President
300 Lakeside Drive, 18th Floor
Oakland, California 94612
Ph: 510-987-9479
E-mail: ellen.switkes@ucop.edu

Dr. Jolinda Traugh
Professor of Biochemistry
3410 Boyce Hall
UC, Riverside
Riverside, California 92521
Ph: 909-787-4239
E-mail: traugh@ucr.edu

Dr. John Vaughn
Executive Officer
Director of Education Policy
Assoc. of American Universities
1 Dupont Circle, Suite 730
Washington, D.C., 20036
Ph: 202-466-5030
E-mail: john_vaughn
@aau.nche.edu

Dr. Frederic Wan
Vice Chancellor of Research
Chancellor's Office
155 Administration, 3175
UC, Irvine
Irvine, California 92717
Ph: 714-824-5809
E-mail: fwan@uci.edu

Dr. Huber Warner
Assistant Deputy Head
Life Sciences Division
Mail Stop 29-100
Berkeley National Laboratory
1 Cyclotron Road
UC, Berkeley
Berkeley, California 94720
Ph: 510-486-6389

Dr. Gabriele Wienhausen
Senior Lecturer
Department of Biology
3080D York Hall, 0355
9500 Gilman Drive
UC, San Diego
La Jolla, California 92093
Ph: 619-534-3018
E-mail: gweinhausen@ucsd.edu

Dr. Renee Williard
Pew Policy Fellow
Center for the Health Professions
1388 Sutter Street, Suite 805
UC, San Francisco
San Francisco, California 94109
Ph: 415-502-5406
E-mail: rlw@itsa.ucsf.edu

Dr. Leslie Wilson
Adjunct Assistant Professor
School of Pharmacy
521 Parnassus Avenue, C156
UC, San Francisco
San Francisco, California 94143
Ph: 415-502-5092

Dr. Keith Wilson
Director, Faculty Relations
Academic Affairs
S115 Medical Sciences Building
512 Parnassus Avenue
UC, San Francisco
San Francisco, California 94143
Ph: 415-476-5742

Dr. Charles B. Wilson
Senior Associate to the President on Medical Affairs
UC Office of the President
300 Lakeside Drive, 9th Floor
Oakland, California 94612
Ph: 510-987-9351

Dr. Ernest M. Wright
Professor and Chair
Department of Physiology
53-317 CHS
10833 Le Conte Avenue
UC, Los Angeles
Los Angeles, California 90095
Ph: 310-825-6905
E-mail: ernest@physiology.medsch.ucla.edu

Cherisa Yarkin
Project Director
Critical Linkages Project
345 Giannini Hall
UC Systemwide Biotech Prgrm
UC, Berkeley
Berkeley, California 94720
Ph: 510-643-0725

Dr. Patricia Zambryski
Prof/Head Graduate Advisor
Department of Plant Biology
221A Koshland Hall
UC, Berkeley
Berkeley, California 94720
Ph: 510-643-9203
E-mail: zambrysk@mendel.berkeley.edu

Dr. Ami Zusman
Coordinator
Graduate Education Planning
Academic Affairs
UC Office of the President
Oakland, California 94612
Ph: 510-987-9511
E-mail: ami.zusman@ucop.edu